Balancing between Trade and Risk

The trade aspects of risk and the risk aspects of trade deserve more systematic and genuine interdisciplinary attention if we are to really understand the global, international and supranational dimensions of risk regulation. This book brings together legal and social science research on risk regulation from across the world to explore risk regulation in a trade context. The chapters individually are worth reading, but it is the set of chapters taken together that offers an interdisciplinary assessment of critical issues in balancing trade and risk. The interdisciplinary collaboration provided in this book is needed to address the balancing act between trade and risk both in empirical and theoretical terms. Although it is obvious that legal, social, cultural and political matters interfere with risk regulation, analyses in which these interferences are adequately considered are lacking. In one way or another, all chapters in this book address the issue of scientific uncertainty, the governance arrangements around expertise or both. Issues such as transparency, trust, legitimacy and precaution also become particularly important given the political, multi-actor and multi-level governance characteristics of balancing trade and risk regulation. This book highlights and examines these concerns, going on to provide a critical assessment of EU regulation of trade and risk from both external and internal perspectives. This book's exploration of trade *versus* risk regulation will be increasingly important to students of law and social sciences as they move to a shared, interdisciplinary understanding.

Marjolein B.A. van Asselt is Professor of Risk Governance, Faculty of Arts and Social Sciences of Maastricht University, and member of the Scientific Council for Government Policy (WRR), The Hague.

Esther Versluis is Associate Professor European Regulatory Governance at the Department of Political Science, Maastricht University.

Ellen Vos is Professor of European Union Law at the Law Faculty of Maastricht University and co-director of the Maastricht Centre for European Law of Maastricht University.

Balancing between Trade and Risk

Integrating legal and social science perspectives

Edited by Marjolein B.A. van Asselt,
Esther Versluis and Ellen Vos

Routledge
Taylor & Francis Group

LONDON AND NEW YORK

earthscan
from Routledge

First published 2013
by Routledge
2 Park Square, Milton Park, Abingdon, Oxfordshire OX14 4RN

Simultaneously published in the USA and Canada
by Routledge
711 Third Avenue, New York, NY 10017

First issued in paperback 2014

Routledge is an imprint of the Taylor & Francis Group, an informa business

British Library Cataloguing in Publication Data
A catalogue record for this book is available from the British Library

Library of Congress Cataloging in Publication Data
Balancing between trade and risk : integrating legal and social science
perspectives / edited by Marjolein B.A. van Asselt, Esther
Versluis, and Ellen Vos.
p. cm.
Includes bibliographical references and index.
1. European Union countries--Commerce. 2. Risk management--European
Union countries. I. Asselt, M. B. A. van II. Versluis, Esther, 1975-
III. Vos, Ellen.
HF3496.5.E85B35 2013
382.094--dc23
2012026326

ISBN: 978-1-84971-361-0 (hbk)
ISBN: 978-1-138-90099-8 (pbk)
ISBN: 978-0-203-10990-8 (ebk)

Typeset in Goudy
by Taylor & Francis Books

Contents

Figures

Tables

Contributors

Marjolein B.A. van Asselt, Chair in Risk Governance at Maastricht University and member of the Scientific Council for Government Policy, the Hague, the Netherlands

Tessa Fox, PhD Researcher, Faculty of Arts and Social Sciences, Maastricht University, the Netherlands

Vessela Hristova, Research Fellow, Institute for European Integration Research, University of Vienna, Austria

Boryana Ivanova, Compliance Analyst, Deutsche Bank Risk Center, Berlin, Germany and Maastricht Graduate School of Governance, the Netherlands, alumna

Anne-May J.P. Janssen, former student and teacher at Maastricht University (the Netherlands) and currently participating in the trainee program of the Dutch government, Ministry of Economic Affairs, Agriculture and Innovation

Jinhee Kim, PhD Researcher, Faculty of Arts and Social Sciences, Maastricht University, the Netherlands

Christoph Klika, PhD Researcher, Faculty of Arts and Social Sciences, Maastricht University, the Netherlands

Renata Motta, Doctoral Researcher at desiguALdades.net, Freie Universität Berlin, Germany

Madjar Navah, Postgraduate Student M.Sc. International Public Policy at University College London, UK

Frank Rodrigues, PhD student at the School of Law, Birkbeck College, University of London, UK

Esther Versluis, Associate Professor European Regulatory Governance, Faculty of Arts and Social Sciences, Maastricht University, the Netherlands

Ellen Vos, Professor of European Union Law, Faculty of Law, Maastricht University, the Netherlands

Maria Weimer, post-doctoral researcher at Maastricht University, the Netherlands

Karolina Zurek, senior researcher in law at the Swedish Institute for European Policy Studies (SIEPS) and postdoctoral researcher at the Centre for Baltic and East European Studies at Södertörn University in Stockholm

Abbreviations

Art.	Article
BBP	Blood-Borne Pathogen
BMELV	Federal Ministry of Food, Agriculture and Consumer Protection
BPA	Bisphenol A
BSE	Bovine spongiform encephalopathy
CA	Competent Authorities
CAP	Common Agricultural Policy
CD	Compact Disc
CEE	Central and Eastern Europe
CFI	Court of First Instance
CHMP	Committee for Medicinal Products for Human Use
CLP	Classification, Labelling and Packaging
CNA	Competent National Authority
CPMP	Committee for Proprietary Medicinal Products
CRM	Carcinogens, Reproductive Toxins and Mutagens
CVMP	Committee for Medicinal Products for Veterinary Use
DEHP	Diethylhexyl Phthalate
DES	Diethylstilbestrol
DG	Directorate General
DIDP	Diisodecyl Phthalate
DINP	Diisononyl Phthalate
DNOP	Di-n-octyl Phthalate
EC	European Commission
EC	European Communities
ECHA	European Chemicals Agency
ECJ	European Court of Justice
EFSA	European Food Safety Authority
EGE	European Group on Ethics of Science and New Technologies
EMA	European Medicines Agency
EU	European Union
EUI	European University Institute
EUR	Euros

FAQ	Frequently Asked Questions
FDA	Food and Drug Administration
FOE	Friends of the Earth
GAIN	Global Agriculture Information Network
GATT	General Agreement on Tariffs and Trade
GDP	Gross Domestic Product
GM	Genetically Modified
GMO	Genetically Modified Organism
HBCDD	Hexabromocyclododecane
IFPRI	International Food Policy Research Institute
IRGC	International Risk Governance Council
JRC	Joint Research Centre
MEP	Member of European Parliament
MSC	Member State Committee
NCAs	National Competent Authorities
NGO	Non-Governmental Organisation
OJ	Official Journal
Para.	Paragraph
PBT	Persistent, Bio-accumulating and Toxic
PPH	Primary Pulmonary Hypertension
PVC	Polyvinyl Chloride
QM	Qualified Majority
QMV	Qualified Majority Voting
RAC	Risk Assessment Committee
REACH	Registration, Evaluation, Authorisation and Restriction of Chemicals
Reg.	Regulation
SANCO	General Directorate for Health and Consumer Protection
SCAN	Scientific Committee on Animal Nutrition
SCF	Scientific Committee on Food
SCFCAH	Standing Committee on the Food Chain and Animal Health
SCNT	Somatic Cell Nuclear Transfer
SCP	Scientific Committee on Plants
SDS	Safety Data Sheet
SEAC	Socio-economic Analysis Committee
SIEPS	Swedish Institute for European Policy Studies
SPS	Sanitary and Phytosanitary Measures
SVHC	Substances of Very High Concern
TBT	Technical Barriers to Trade
TDI	Tolerable Daily Intake
TEC	Treaty of the European Community
TFEU	Treaty on the Functioning of the European Union
UK	United Kingdom
UN	United Nations

US	United States
USA	United States of America
VMD	Veterinary Medicines Directorate
WRR	Wetenschappelijke Raad voor het Regeringsbeleid (Dutch Scientific Council for Government Policy)
WTO	World Trade Organization

Introduction

Tessa Fox, Marjolein B.A. van Asselt, Esther Versluis and Ellen Vos[1]

Free Trade Versus Protection of Health and the Environment?

Every day products and technologies involving risks for public health and the environment are traded, at a local, regional, European or international level. This entails that regulators are challenged with the dilemma of allowing free trade and protecting non-trade concerns, such as human health and safety and the environment, which potentially hinder trade. The globalization of trade, in particular under the WTO rules, to a great extent dictates national and EU trade rules as well as the protection of human health and safety and the environment. This forces both national and EU regulators to continuously justify any deviation from free trade in favour of the protection of non-trade interests, in terms of science and scientific evidence. Therefore, trade and non-trade issues are increasingly getting intermingled, as often non-trade, science-based arguments lead to hindering free trade. Moreover, when deciding about protecting non-trade concerns, the question arises as to which level of protection should be set by the regulator, and therefore also, what level of protection and precaution are both acceptable and consistent with the EU and/or international rules.

This highlights the complex tasks for regulators who must observe the EU or international agreements while at the same time take account of domestic (within the European Union) responsibilities and concerns. How to organize the balancing of risk and trade in a system of multi-level governance is a key question. This book aims to provide new insights and reveal how this is dealt with currently. It will identify current shortcomings and make suggestions for change.

The research compiled in this book can be characterized as inter-disciplinary study of risk regulation practices in a trade context. The book brings together, confronts and integrates input and perspectives from legal and social science research on risk regulation. In doing so, it illustrates the added value and the critical potential of collaboration between legal scholars and social scientists. So in terms of agenda-setting, with this book we not only aim to increase the scholarly interest in the trade aspects of risk regulation and the risk dimensions of international trade, but we also attempt to endorse interdisciplinary research between legal scholars and social scientists on risk and trade issues.

Limited Scholarly Attention For Trade in the Risk Community

Are practices in risk regulation currently studied in a trade perspective? In other words, to what extent are the trade dimensions of risk regulation recognized in the scholarship of risk?[2] A search of current issues[3] (2010–2012) of leading risk journals[4] – *Journal of Risk Research* and *Risk Analysis* – and a new journal – the *European Journal of Risk Regulation* – reveals that scholarly attention exposed by the 'risk community' is rather limited (Table I.1). We found 24 papers addressing risk regulation in a trade perspective, out of nearly 800 papers published in these risk journals in the searched period. Thus, only 3 per cent of the papers address the trade dimensions of risk regulation.

Most papers that recognize the trade dimension of risk regulation address the diversity of, and inconsistency between, regulatory frameworks, and the tensions between the EU and the WTO supranational frameworks in particular. This academic debate can be followed in the contributions of Arcuri, Gruszczynski and Herwig (2010/2) (2010/3), Flett (2010) and Gruszczynski (2011/1) (2011/3) in the *European Journal of Risk Regulation*. Some papers discuss a particular domain of regulation, such as Internet-based trade (Littler 2011), trade and nanotechnology (Jaspers 2010; Van Broekhuizen and Reijnders 2011), trade and biotechnology (Varela 2010) and chemicals regulation in the EU (Nordlander *et al.* 2010).

Table I.1 Risk-trade literature overview – Articles including the word 'trade' in title or abstract in *European Journal of Risk Regulation, Journal of Risk Research and Risk Analysis*, 2010, 2012. The total number of articles is 24.* Articles found through snowballing (references in the papers found in the systematic search).

	EU – WTO	Other (Case studies)
Variety of regulatory frameworks	Arcuri *et al.* (2010/2) (2010/3), Hristova (2010), Macrae (2011), Nordlander *et al.* (2010), Varela (2010), Van Zeben (2010), Weimer (2010)	Alemanno (2010), Arcuri *et al.* (2010/4), Gruszczynski (2011/1) and (2011/3), Jaspers (2010), Littler (2011), McGrady (2011), Vadi (2011), Sánchez-Vizcaíno *et al.* (2010/5) and (2011/12)
Other (Trade and the precautionary principle)	Flett (2010), Van der Meulen (2010), Wintle and Cleeland (2012), Gabbi (2011)*, Rogers (2011)*, Anders and Schmidt (2011)* and Zandvoort (2011)*	Jung *et al.* (2009), Mohan and Aggarwal (2009), Van Broekhuizen and Reijnders (2011)

The European Union is a leading trade block and a key actor in the international market. As a consequence, the EU is involved in many of the controversies surrounding trade and risk and the significant cases for the WTO in particular. This explains the interest for the EU and the WTO in the papers addressing risk regulation in a trade perspective, such as Arcuri *et al.* (2010/2) (2010/3), Hristova (2010), Macrae (2011), Nordlander *et al.* (2010), Varela (2010), Van Zeben (2010) and Weimer (2010).

Concerning the role of scientific risk assessment and of scientific expertise, the recurring discussion on the problems of the functional separation between risk assessment and risk management echoes. The trade perspective, furthermore, reaffirms or even highlights the relevance of various interpretations of scientific authority, evidence and uncertainty to this balancing act. Decision-makers try to resolve the dilemma of the governance of free trade, consumer safety and environmental health by resorting to scientific expertise, also in situations of scientific uncertainty about causal mechanisms, dose–effect relationships and the nature of the risk, the order of magnitude, and the likelihood of occurrence. It can be argued that, to a certain extent, critical questions about regulatory inconsistency boil down to the frictions arising in governance arrangements around expertise and dealing with uncertainty.

Also, the tendency towards regulatory diversity, inconsistency and tensions can be recognized as an important emerging theme concerning risk regulation in a trade context. Both Van Zeben and Zandvoort discuss legitimacy and 'legal liability' of risk decisions taken in supra-national bodies. Van Zeben (2010) discusses legal liability under the European Emissions Trading Scheme, whereas Zandvoort (2011) discusses legal liability for technological risks in international environmental law.

The Complexity of Risk Regulation in a Trade Context

Regulating consumer safety and environmental health in an international trade context is complex. This is due to a variety of factors, such as diversity, inconsistencies and tensions pertaining to political cultures (see also Jasanoff 2005; Wiener *et al.* 2011), differences in regulatory frameworks and in standards for risk assessment in addition to definitions, interpretation and application of the precautionary principle, as well as questions of legitimacy.

Policy makers are confronted with the need to regulate a host of products as diverse as feedstuffs, food products and additives, packaging, clothes, toys, cosmetics, chemicals and pharmaceuticals while also facing scientific uncertainty regarding health or environmental effects of these products (see for example, Varela 2010; Stampfli *et al.* 2010; Fox *et al.* 2011; Van Asselt and Vos 2006, 2008; Löfstedt 2009 and 2011). The rapid development of new technologies and products makes it difficult, if not impossible, to keep track of the possible environmental and health impacts. Research is always lagging behind and effects may materialise in the long run. This phenomenon has been referred to as the latency lacuna (Harremoes 2002). How should we take decisions to prevent or control possible outbreaks of food-borne illnesses, significant long-term health effects or (acute or latent) environmental effects from the use of specific products or technologies? How should we deal with the uncertainty of such risks and thus also, the underlying knowledge problems and associated societal ambiguity, in view of trade interests and concerns?

Trading around the globe adds substantial momentum to the regulatory challenge because potential risks associated with the traded products also spread, or even increase, as the vulnerability – or more positively phrased: the resilience – of different societies with regard to particular risks may also differ. How should we deal with the transboundary dimension of risk management (Löfstedt 2001) in a multi-level governance world?

Decision-makers are often confronted with risk issues in the context of trade. Crises such as the BSE crisis (1996–2000), E. coli outbreaks (2004–2010) and incidents such as those concerning salmonella in peanut butter (2007), all involved trade products including meat, cattle and other food products. Such crises have arguably heightened the pressure on regulators and decreased public trust in decision-makers (Beck and Kropp 2011; Anders and Schmidt 2011; Vos and Wendler 2006). The risk aspects of trade also captures media attention (Löfstedt 2009 and 2011; Fox et al. 2011), which in its turn impacts on public awareness and demands for consumer safety and environmental health (Kasperson et al. 1988; Pidgeon 2003). Thus, risk regulation – which is already complex due to uncertainty and the transboundary and multi-level nature of risk regulation – is taking place in a context characterized as 'post-trust' (Löfstedt 2005). In such contexts, innovation, technology, industrial activities and associated waste issues are more easily associated with risk. At the same time zero-risk statements or other forms of assurance by authorities are the subject of distrust. Hence sufficient trust in and support for risk regulation has to be gained actively.

From the current state of affairs it can be concluded that risk regulation in a trade context is complex for a number of reasons ranging from diversity in political cultures and regulatory frameworks, the uncertain and transboundary nature of the risk, to societal pressures and demands. How to balance trade and risk is not only an important question but a highly complex challenge for regulators.

This Book

This book aims to offer some scholarly insights into these questions. Interdisciplinary collaboration is needed to address the trade *versus* risk balancing act both in empirical and theoretical terms. The trade aspects of risk and the risk aspects of trade deserve more systematic and genuine interdisciplinary attention if we are to really understand the global, international and supranational dimensions of risk regulation. This book aims to increase the scholarly interest in systematic, symmetric and interdisciplinary research on risk regulation in a trade context. It also aims to provide a critical assessment of the EU regulation of trade and risk both from external and internal perspectives. Although it is obvious that legal, social, cultural and political matters interfere, analyses in which these interferences are adequately considered are lacking. This volume brings together and synthesizes input from legal and social sciences, providing complementary views on risk regulation in a trade context.

From a social science perspective, legal science helps provide a more adequate understanding of legal realities and a less utopian view on issues such as participation and precaution as the legal discipline requires answering questions pertaining to process design and regulatory frameworks. From a legal science perspective, social sciences help to question the often utopian view of the role of experts, putting them on a pedestal and considering scientific advice as simple facts. As the social sciences stressing the social nature of knowledge construction, this forces legal discipline to question the legal position of experts in regulation. Consequently, interdisciplinary collaboration between social and legal scientists allows for improved examination of governance arrangements, regulatory principles and the challenges of dealing with uncertainty.

In one way or another, all chapters in this book address the issue of scientific uncertainty, the governance arrangements around expertise or both. This volume indicates the need to rethink important issues that lie at the heart of risk regulation in a trade context, such as transparency, trust, legitimacy and precaution. The political, the multi-actor and the multi-level governance characteristics of the balancing act are highlighted and examined. This book also reveals asymmetries in the factors considered and weighted in the regulation of risk and trade. Finally, the problem of accommodating diversity – both in cultural and legal terms – when balancing trade and risk, is identified and investigated.

The various chapters provide original insights into the non-trade dimensions of what are usually considered to be trade dossiers and the trade dimensions of what are often treated as risk dossiers, ranging from genetically modified organisms and cloned animals to chemicals in child products and nanotechnology. Through interdisciplinary analysis, new issues are highlighted, provoking conclusions are drawn and a demanding and policy relevant research agenda is sketched, with the aim of inspiring both academics and actors actually involved in balancing risk and trade.

There are three parts to this book, each of which comprises three or four complementary chapters: (I) external perspectives on EU risk regulation in a global trade context, in which part the implications of EU regulation on international trade are considered; (II) internal perspectives on the regulation of risk and trade in the European Union, in which part the issue of accommodating different types of diversity within the European Union are addressed, and; (III) risk governance, in which policy options are evaluated.

Part I – EU Risk Regulation in a Trade Context: External Perspectives

This part contains four chapters that analyse the regulation of innovative food in the EU. However, they do so from an external perspective as they reflect on EU risk regulation in view of the implications on international trade. In the first chapter, Zurek analyses EU food regulation during the last two rounds of enlargement of the Union. Weimer analyses the proposed amendment to the Novel Food Regulation in view of the treatment of cloned food. Her chapter explores the implications the European approach has on trade with third countries. The conflict between the EU and the USA on GMOs are central in the analysis of both Motta and Ivanova and Van Asselt.

Zurek analyses the extent to which the concerns of the 12 new Member States regarding the food market were taken on board during the last two enlargement rounds (2004 and 2007). Informed by an evolutionary analysis of EU food regulation, Zurek argues that with the increased diversity of the EU and the growing interdependence on the global arena, the way in which the EU regulates food requires more flexibility in order to take account of a broader range of concerns. However, according to Zurek the political will to allow a more inclusive approach in EU food regulation is lacking. It is a matter of whether the socio-economic implications of food are considered irrelevant or too problematic.

In her chapter on the difficulties of regulating cloned food in the EU (the Novel Food Regulation), Weimer discusses three dimensions that complicate regulation in this area: (1) the uncertainty of the risks; (2) the salience of ethical values; and (3) the strong international trade dimension. At stake is whether the EU can create legal frameworks that ensure the right balance between risk control and the accommodation of societal concerns on the one hand, and the promotion of technological innovation and global free trade on the other.

Motta analyses the WTO dispute between the USA, Canada and Argentina and the EU arising over the EU's emphasis on the existence of risks associated with GMO products: the *EC-Biotech case*. Motta argues that the countries frame the conflict differently by analysing how both sides of the controversy use political, scientific and economic arguments to dispute which interpretation of international law prevails, in order to judge the European policy on GMOs. This chapter is an attempt to bridge legal studies and social sciences by using conceptual tools from the social sciences in the analysis and explanation of phenomena usually studied under the auspices of law.

Ivanova and Van Asselt too discuss the *EC-Biotech case*. They focus on whether it matters for such trade conflicts how uncertainty is dealt with in the risk assessments, and if so, how. To that end, they analyse the reasoning of the WTO panel pertaining to uncertainty, risk assessment and precautionary measures. Ivanova and Van Asselt conclude that uncertainty intolerant risk assessments foreclose the space for risk managers to adopt precautionary strategies in dealing with uncertain risks. So how uncertainty is dealt with in trade-risk conflicts is highly politically relevant.

Part II – EU Risk Regulation in a Trade Context: Internal Perspectives

The three chapters in this part examine how and why it is difficult to deal with (uncertain) risks pertaining to traded products in a heterogeneous European Union. This part deals with accommodating different types of diversity. The EU includes many different countries with diverging traditions and backgrounds and hence also different legal and cultural regimes. As a consequence, trying to adopt a common regulatory approach to risk is by definition a complex and complicated endeavour. Hristova touches upon the question of how to accommodate national diversity in the EU GMO authorization procedure. Navah, Versluis and Van Asselt investigate political diversity by mapping Member States' voting behaviour in Comitology and the Council of Ministers on GMO cases. Finally, Fox examines the importance to acknowledge the diversity of interests in balancing trade and risks. She analyses how health and trade concerns are accommodated within the EU by exploring the regulation of the chemical Bisphenol A in baby products.

Hristova shows that a paradox exists at the heart of European integration: integration seems to require that, at some times and in some ways, national diversity is accommodated. The EU is united through diversity. She explores the fault-line between the pressures for unity and the counteracting pressure for maintaining variety at the local level of specificities. She explores the mechanisms through which the regulatory framework can reconcile as well as accommodate diversity. Informed by her analysis of the EU regulation of agricultural biodiversity, this chapter can be considered an important contribution to the risk assessment *versus* risk management debate.

Navah, Versluis and Van Asselt contribute to the current debate on the regulation of trade on GMOs to and within the EU, by providing insight into the politics of European decision-making. By evaluating 18 decision-making procedures in the Standing Committee on the Food Chain and Animal Health (SCFCAH) and in the Council of the EU, they analyse the political diversity within the EU and examine the extent to which voting patterns can be understood in terms of classical political science categorizations of the domestic political landscape. This analysis helps to better understand the political dimension of the difficulty to reach a qualified majority when deciding on the market approval of specific GMOs.

The contribution by Fox analyses whether and how the EU should regulate the use of Bisphenol A (BPA) in baby products, taking into account the lack of certainty and possible trade implications. This chapter provides an understanding of how the European Union approaches regulation in the face of uncertainty surrounding risks. On the basis of the current regulatory tools in place and the ban on BPA in baby bottles, which was invoked by the EU in March 2011, the contribution presents four scenarios, each depicting other possible regulatory futures for Bisphenol A in the EU.

Part III – EU Risk Governance in a Trade Context

In the last part of this volume several policy options that might provide some answers as to how to regulate or govern risks are evaluated and discussed. Particularly when dealing with *uncertain* risks, reliance on science becomes problematic: what do strategies and approaches – such as agencies, the precautionary principle and participation – have to offer here? Kim, Klika and Versluis analyse the role of agencies in regulating risks in a trade context, by comparing the authorization procedures of three different EU agencies. Janssen and Van Asselt touch upon the role of the precautionary principle in case law pertaining to regulating risks. Rodrigues discusses the issue of participation. He argues that when dealing with uncertain risks, participation begins where science ends.

Kim, Klika and Versluis examine agencies, currently a dominant governance arrangement in the regulation of trade and risk. They investigate what role three agencies – the EMA, EFSA and ECHA – actually play in the process of EU risk regulation of traded products in general and medicinal products, GMOs and chemical products in particular. The authors argue that European agencies seem to function as risk assessors, risk managers and risk communicators, albeit that there are some differences between the agencies. Although portrayed as risk assessors, the EMA and EFSA seem to act as *de facto* risk regulators, whereas the ECHA's role is less clear due to the fact that it is a relatively new agency. However, based on the similarity in their implementing procedures, it is to be expected that their role in risk management will also be considerable.

In their analysis of the role of the precautionary principle in the EU court, Janssen and Van Asselt first summarize the milestone *Pfizer* case, on virginiamycin used as growth promoter in feed. Then, three post-Pfizer cases are discussed. Various patterns and inconsistencies in dealing with uncertainty are identified and analysed, ranging from issues pertaining to risk assessment, demonstrating a disregard for the temporary nature of precautionary measures and inconsistencies in the way in which the Court reviews similar cases. The authors argue that their findings suggest the *Pfizer* case, which is broadly considered a problematic application of the precautionary principle, set a precedent. Janssen and Van Asselt conclude with some options to improve the approach to the precautionary principle.

In his chapter, Rodrigues delves into the regulatory dilemma posed by uncertainty. When regulating uncertain risks, reliance on scientific input becomes questionable. A current response is to advocate a more pluralistic model involving participation as a means to achieve both substantive regulatory decisions and a guarantee of their legitimacy. Participation is expected to deliver more transparent and legitimate decisions. Rodrigues claims there are reasons for scepticism as to these assumptions.

Towards Interdisciplinary Synthesis

Taken together, this interdisciplinary volume covers the development of legal frameworks, the process and politics of balancing trade and risk, the role of science in risk regulation, the dynamics between the legal and policy spheres and the societal embeddedness of law and policymaking. Hence, the different phases and different dimensions of risk regulation in the broadest possible meaning are discussed in this volume, both on the level of international trade and with regard to risk regulation in the European Union more in particular.

The chapters individually are worth reading, but it is the set of papers taken together that offers an interdisciplinary assessment of critical issues in balancing trade and risk. The chapters are complementary and mutually challenging. In the synthesis of the book (Chapter 11) we compare and contrast the observations, insights and thoughts in the various chapters. In doing so it provides a richer and more substantiated view of the social reality around trade and risk and it brings new issues to the fore that emerge only from the synthesis of the various contributions.

This volume demonstrates how young scholars respond to the plea for interdisciplinary research between law and social science. Notwithstanding the differences in research approach, the various studies covered in this volume result in relevant observations, insights and thoughts that deserve to be talked about. But it also provides a basis for reflection on the future direction of interdisciplinary law–social science research. Furthermore, in Chapter 11 we will discuss what can be learned from this volume on European risk regulation in a trade perspective. The chapters taken as a set invite reflection on regulatory arrangements around expertise and the broader issue of depoliticizing and 'scienticizing' decision-making on trade and risk. Therefore, we also discuss what can be learned from this volume about regulatory arrangements around expertise.

We think that the insights gained in this volume demonstrate that integrating legal and social science perspectives seriously enhances our understanding of critical dimensions of balancing trade and risk, and that it is thus definitely worth further pursuit, notwithstanding the required and sometimes demanding efforts to mutually understand different research traditions, bodies of knowledge and research approaches.

Notes

1 We greatly acknowledge the assistance of Tonje J. Espeland, student at the Faculty of Arts and Social Sciences of Maastricht University, who helped us enormously by doing research on the scholarly attention to the trade dimension of risk regulation. We would like to express our sincere thanks to her.

2 The question could also be asked whether, and to what extent, legal and trade scholars address the risk dimension of trade. In the legal literature (books and articles) risk regulation is discussed in a trade context, e.g. with references to the World Trade Organization (WTO) (see, for example, Alemanno 2007; Bermann

and Mavroides 2006; Button 2004; Gruszczynski 2011; Prevost 2009). In this book, however, we 'limit' ourselves to integrating input and perspectives from legal scholars and social scientists that are, more or less, part of the multi-disciplinary risk community and share an interest in the trade dimension of risk regulation. We thus aim to set an agenda regarding the issue of trade in this risk community. A full multidisciplinary review of all relevant literature on risk and trade is beyond the scope of this introductory chapter. Of course, in the upcoming chapters references are made to the legal body of literature on trade and risk.

3 We did a search with 'trade' as search item for title and abstract and we scanned all the issues on title and abstract level to find other papers that prove to be interesting voices to the risk and trade debate. Examples of the results include Rogers (2011); Zandvoort (2011); Sánchez-Vizcaíno *et al.* (2010); Gabbi (2010); Vareman and Persson (2010) and Anders and Schmidt (2011).

4 Journals can be systematically searched and as media for peer-reviewed publications, they provide a relevant and researchable site for examining the current level of scholarly interest in risk regulation in a trade perspective. However, it should be noted that especially in the legal sciences (but also in the social sciences) books are still highly important publication devices. Various risk scholars have also written PhD theses or books on risk regulation in a trade context, such as Gruszczynski (2010) and Alemanno (2007). The results of the journal search can be considered an indicator for the relative level of scholarly attention in the risk community, which was the aim of the search, but it cannot be considered a comprehensive overview of relevant risk research.

References

Alemanno, A. (2007) *Trade in Food: Regulatory and Judicial Approaches in the EC and the WTO*, London: Cameron May.

——(2010) 'The First GMO Case in Front of the US Supreme Court: To Lift or Not to Lift the Alfalfa Planting Ban?', *European Journal of Risk Regulation*, 1(2): 152–53.

Anders, S. and Schmidt, C. (2011) 'The International Quest for an Integrated Approach to Microbial Food-borne Risk Prioritization: Where do we Stand?', *Journal of Risk Research*, 14(2): 215–39.

Arcuri, A., Gruszczynski, L. and Herwig, A. (2010a) 'Global Governance of Risks – WTO, Codex Alimentarius and Private Standards – Report on the SRA-Europe 19th Annual Conference', *European Journal of Risk Regulation*, 1(3): 285–87.

——(2010b) 'Independence of Experts and Standards for Evaluation of Scientific Evidence under the SPS Agreement – New Directions in the SPS Case Law', *European Journal of Risk Regulation*, 1(2): 183–88.

——(2010c) 'Risk Apples Again? Australia – Measures Affecting the Importation of Apples from New Zealand', *European Journal of Risk Regulation*, 1(4): 437–43.

Beck, G. and Kropp, C. (2011) 'Infrastructures of Risk: A Mapping Approach towards Controversies on Risk', *Journal of Risk Research*, 14(1): 1–16.

Bermann, G. and Mavroides, P. (eds) (2006) *Trade and Human Health and Safety*, Cambridge: Cambridge University.

Button, C. (2004) *The Power to Protect. Trade, Health and the WTO*, Oxford: Hart Publishing.

Flett, J. (2010) 'If in Doubt, Leave it Out? EU Precaution in WTO Regulatory Space', *European Journal of Risk Regulation*, 1(1): 20–30.

Fox, T., Versluis, E. and Van Asselt, M.B.A. (2011) 'Regulating the Use of Bisphenol A in the European Union', *European Journal of Risk Regulation*, 2(1): 21–35.

Gabbi, S. (2011) 'Independent Scientific Advice: Comparing Policies on Conflicts of Interest in the EU and the US', *European Journal of Risk Regulation*, 2(2): 213–26.

Gruszcynski, L. (2010) *Regulating Health and Environmental Risks under WTO Law*, Oxford, UK: Oxford University Press.

——(2011a) 'How Deep should we go? Searching for an Appropriate Standard of Review in the SPS Cases', *European Journal of Risk Regulation*, 2(1): 111–14.

——(2011b) 'Trade, Investment and Risk: United States: Certain Measures Affecting Imports of Poultry from China – Just another SPS Case?', *European Journal of Risk Regulation*, 2(3): 432–37.

Harremoes, P. (2002) *The Precautionary Principle in the 20th Century: Late Lessons from Early Warnings*, London: Earthscan.

Hristova, V. (2010) 'Biotechnology: Recent Developments in EU Biotech Regulation: A Possible Solution to the Deadlock on Authorizations of GM Crops?', *European Journal of Risk Regulation*, 1(2): 151–52.

Jasanoff, S. (2005) *Designs on Nature: Science and democracy in Europe and the United States*, Princeton: Princeton University Press.

Jaspers, N. (2010) 'Nanotechnology: How to Avoid International Trade Conflicts', *European Journal of Risk Regulation*, 1(2): 167–73.

Jung, J., Santos, J.R. and Haimes, Y.Y. (2009) 'International Trade Inoperability Input – Output Model (IT-IIM)', *Theory and Application Risk Analysis*, 29(1): 137–54.

Kasperson, R., Renn, O., Slovic, P. *et al.* (1988) 'The Social Amplification of Risk: A Conceptual Framework', *Risk Analysis*, 8: 177–87.

Linnerooth-Bayer, J., Löfstedt, R.E. and Sjöstedt, G. (eds) (2001) *Transboundary Risk Management*, London: Earthscan.

Littler, A. (2011) 'Internet-Based Trade and the Court of Justice: Different Sector, Different Attitude', *European Journal of Risk Regulation*, 2(1): 78–84.

Löfstedt, R.E. (2005) *Risk Management in Post-trust Societies*, Hampshire and New York: Palgrave Macmillan.

——(2009) 'Risk Communication and the FSA: The Food Colourings Case', *Journal of Risk Research*, 12(5): 537–57.

——(2011) 'Risk versus Hazard – How to Regulate in the 21st Century', *European Journal of Risk Regulation*, 2(2): 149–68.

Macrae, D. (2011) 'Standards for Risk Assessment of Standards: How the International Community is Starting to Address the Risk of the Wrong Standards', *Journal of Risk Research*, 14(8): 933–42.

McGrady, B. (2011) 'Panel Report US – Clove Cigarettes', *European Journal of Risk Regulation*, 2(4): 600–6.

Mohan, R.M.P. and Aggarwal, V. (2009) 'Spent Fuel Management in India', *Journal of Risk Research*, 12(7–8): 955–67.

Nordlander, K., Simon, C.M. and Pearson, H. (2010) 'Hazard v. Risk in EU Chemicals Regulation', *European Journal of Risk Regulation*, 1(3): 239–50.

Pidgeon, N. (2003) *The Social Amplification of Risk*, Cambridge: Cambridge University Press.

Prevost, D. (2009) *Balancing Trade and Health in the SPS Agreement: The Development Dimension*, Nijmegen: Wolf Legal Publishers.

Rogers, M.D. (2011) 'Risk Management and the Record of the Precautionary Principle in EU Case Law', *Journal of Risk Research*, 14(4): 467–84.

Sánchez-Vizcaíno, F., Perez, A., Lainez, M. and Sánchez-Vizcaíno, J.M. (2010) 'A Quantitative Assessment of the Risk for Highly Pathogenic Avian Influenza Introduction into Spain via Legal Trade of Live Poultry', *Risk Analysis*, 30(5): 798–807.

Sánchez-Vizcaíno, F., Perez, A., Martínez-López, B. and Sánchez-Vizcaíno, J.M. (2011) 'Comparative Assessment of Analytical Approaches to Quantify the Risk for Introduction of Rare Animal Diseases: The Example of Avian Influenza in Spain', *Risk Analysis*, doi: 10.1111/j.1539–6924.2011.01744.x.

Stampfli, N., Siegrist, M. and Kastenholz, H. (2010) 'Acceptance of Nanotechnology in Food and Food Packaging: A Path Model Analyses', *Journal of Risk Research*, 13(3): 353–65.

Vadi, V.S. (2011) 'Trade, Investment and Risk: Overlapping Regulatory Spaces: The Architecture of NAFTA Chapter 11 and the Regulation of Toxic Chemicals', *European Journal of Risk Regulation*, 2(4): 586–90.

Van Asselt, M.B.A. and Vos, E. (2006) 'The Precautionary Principle and the Uncertainty Paradox', *Journal of Risk Research*, 9(4): 313–36.

——(2008) 'Wrestling with Uncertain Risks: EU Regulation of GMOs and the Uncertainty Paradox', *Journal of Risk Research*, 11(1/2): 281–300.

Van Broekhuizen, P. and Reijnders, L. (2011) 'Building Blocks for a Precautionary Approach to the use of Nanomaterials: Positions Taken by Trade Unions and Environmental NGOs in the European Nanotechnologies Debate', *Risk Analysis*, 31(10): 1646–57.

Van der Meulen, B. (2010) 'Prior Authorisation Schemes: Trade Barriers in Need of Scientific Justification', *European Journal of Risk Regulation*, 1(4): 465–71.

Van Zeben, J. (2010) 'Possibilities for Locu Standi and Non-Contractual Damages for Private Parties under the European Emissions Trading Scheme', *European Journal of Risk Regulation*, 1(4): 473–78.

Varela, J.C. (2010) 'Biotechnology', *European Journal of Risk Regulation*, 1(1): 63–71.

Vos, E.I.L. and Wendler, F.A. (2006) 'Food Safety Regulation at the EU Level', in Vos, E. and Wendler, F. (eds), *Food Safety Regulation in Europe. A Comparative Institutional Analysis*, Antwerp-Oxford: Intersentia, 65–138.

Weimer, M. (2010) 'The Regulatory Challenge of Animal Cloning for Food – The Risks of Risk Regulation in the EU', *European Journal of Risk Regulation*, 1(1): 31–9.

Wiener, J.B., Rogers, M.D., Hammitt, J.K. and Sand, P.H. (2011) *The Reality of Precaution. Comparing Risk Regulation in the United States and Europe*, London: RFF Press.

Wintle, B.C. and Cleeland, B. (2012) 'Interpreting Risk in International Trade', *Journal of Risk Research*, 15(3): 293–312.

Zandvoort, H. (2011) 'Evaluation of Legal Liability for Technological Risks in View of Requirements for Peaceful Coexistence and Progress', *Risk Analysis*, 31(6): 969–82.

Part I

The European Union in Context

1   Regulating Food Trade in the Enlarged European Union

Karolina Zurek

Introduction

European food regulation, perceived in the optic of modern risk regulation, has attracted a considerable volume of research in the past years. Hence, its analysis has concentrated mainly on the various aspects of the problem of regulating risks, while the trade and market aspect of food regulation – equally important and tightly interrelated with this problem – has attracted much less interest and scrutiny. This contribution analyses the combination of these two aspects of food regulation and emphasises the significance of its dual character in the Union's new, post-enlargement circumstances.

The current form of food regulation in Europe remains largely shaped by the post-BSE crisis thinking. Thus it is targeted more to respond to crises rather than to accommodate a wider range of regulatory concerns, the further diversification of which results from the accession of the 12 new Member States in 2004 and 2007. In such a diversified market, it is increasingly difficult for regulation to respond to needs and expectations of all localities and societies involved, resulting in its gradual disembedding (Polanyi 2001). The regulatory scheme thus becomes unable to adequately confront the new challenges at stake and fails to fulfil its objectives.

This contribution explores how 'free' and how 'common' the food market has been for the new Member States, and to what extent their diverse concerns are reflected in regulatory decisions. It is argued here that, in the face of increased heterogeneity as a result of the last two enlargements as well as the growing demands of global trade regulation, reorientation of the European Union's regulatory policy should be considered in order to allow for the inclusion of a wider set of socio-economic concerns in the decision-making processes, hence increasing its embeddedness, so as to respond better to the needs of the larger European market and society. Following Polanyi's thinking, it is argued that socio-economic implications of market regulation should no longer be disregarded in the increasingly diversified European Union.

A number of important regulatory concerns seem to lack appropriate consideration, a lack that is aggravated by the post-enlargement increase in diversity. Questions of food quality, sustainable development, and social and economic relevance of the 'Europeanised' policy choices should be mentioned here. If EU food regulation continues to overlook such concerns, significant and far-reaching side effects may result, such as loss of diversity (in both cultural and biological sense), regulatory depreciation of the value of food quality as well as the mismanagement of the problem of obesity.

The first part of this contribution is devoted to a brief introductory ana-lysis of EU food regulation, illustrating the changing objectives and patterns of regulation over time in response to certain external or internal factors. The second part concentrates on the most influential factors currently affecting the content and functioning of EU regulation, namely the challenges of globalization and enlargement, with particular emphasis on the latter. The analysis will aim to show how these developments confront EU food reg-ulation with new problems and concerns, and how they might be addressed. A number of recent internal developments in EU risk regulation are briefly presented to illustrate new approaches to multidimensional concerns in consumer sensitive market areas. Finally, the conclusions reflect on the necessity and practicalities of balancing the current regulatory framework in order to respond better to these new challenges.

The Evolution of EU Food Regulation

With no direct reference to food in the original text of the Treaty, the standing of food in the larger scheme of the EU policies was rather coin-cidental and unstable. Throughout its development during the last couple of decades, food regulation has tended to follow mainstream governance methods, currently *à la mode* in the broader common market regulation. Generally speaking, to a large extent, it has shared the evolutionary path of the Internal Market development, with its reformatory triumphs as well as its problems and mistakes. A number of important features of food regulation, however, make it particularly interesting.

Firstly, the dual character of a foodstuff as a product on the one hand, and an agricultural product on the other, meant that at an early stage of regulation, foodstuffs fell under two different regulatory regimes, namely that for the free movement of goods and that for agriculture. Substantively, these regimes treated the social aspects of regulation differently and failed to balance regulatory approaches according to the different types of foodstuffs concerned. This duality left an important legacy, exemplifying the broader phenomenon in EU market regulation known as 'decoupling' (Scharpf 1999).

Secondly, the complexity and variety of foodstuffs explains the development of food regulation under the principle of the free movement of goods (Lister 1992). Food regulation initially followed the prescriptive, total-harmonisation methodology, producing vertical, recipe-type laws for various categories of products. Faced with the sheer variety of food and production methods, however, the EU legislature had to reconsider this practice, and shifted towards more horizontal harmonisation techniques, approaching problems rather than products.[1] Hence, the regulatory emphasis moved from concentrating on chocolate to focusing on hygiene or labelling practices. Additionally, after the *Cassis de Dijon*[2] judgment of the European Court of Justice, the principle of mutual recognition guided regulatory reform towards concentrating more on essential requirements rather than meticulous technical details (Nikolaidis 2005; Weiler 2005). Following this rationale, the New Approach philosophy was applied to food regulation in the broader Internal Market reform. The New Approach to Technical Harmonisation and Standardisation[3] foresaw that only minimum requirements would be covered by harmonising legislation, while detailed technical rules would be formulated with the participation of technical experts and stakeholders in a form of standards. This integrated market regulatory practice was presented as an attractive alternative to market governance, promising better effectiveness and responsiveness to the needs of the market (Pelkmans 1987; Joerges *et al.* 1999; Schepel 2005). Standardisation opened up the dialogue between various market participants and showed new ways of approaching complex market sectors, under the guidance of specific scientific expertise (Sullivan 1983; Schepel 2005). Food was among those sectors where complexity and multiplicity of regulatory issues justified standardisation in the interests of efficiency as well as the possibility of guaranteeing unified standards for foodstuffs throughout the entire Internal Market.

Thirdly, food is directly linked with the life and health choices of consumers. The following phase of the development of EU food regulation saw the management of food risks handed over to the mechanisms of comitology. As a system of governance, comitology offered a platform for deliberation as well as inclusion of experts and stakeholders. Different types of committees covered different aspects of the risk regulatory process, allowing for the inclusion of scientific expertise as well as various stakeholders' concerns. The functioning of the system, however, faced serious criticism following the outbreak of the BSE pandemic (Chambers 1999; Krapohl 2003). The need to reconsider the fragmented approach to food regulation became apparent, and a 'total' reform was embarked upon in order both to coherently address the entire spectrum of concerns and to rebuild the consumer confidence in the EU. The BSE crisis therefore, marked the 'before' and the 'after' in EU food regulation (Neyer 2000; Little 2001).

The post-BSE reform culminated in Regulation 178/2002,[4] which governs the entire food chain 'from farm to fork' (Vos 2000; Millstone and van Zwanenberg 2001; Vincent 2004). It concentrated mostly on the areas that were challenged by the crisis and needed immediate improvement in order to avoid any repeat mistakes. Creation of the European Food Safety Authority to provide professional and sound scientific expertise for the sake of European risk regulation was a key aspect of the reform. Scientists and stakeholders were involved respectively in separate risk assessment and risk management phases of the regulatory procedure, and the role of science in the decision-making on food was firmly reconfirmed.

These reforms did not live up to the promise of tackling the totality of concerns. Although it dealt extensively with the risk assessment part of the process, emphasising the *safety* aspect of food, other important features were left outside the direct scope of the reform. Paradoxically therefore, partiality in the post-BSE food law reform means that while we might be better prepared for the *unexpected*, it is doubtful whether we are any better prepared for the *expected*. What if the risks connected with a potential outbreak of some sort of crisis are, in fact, outweighed by risks arising due to legal and otherwise safe everyday practices? In terms of food *quality*: while quality to certain extent translates to *safety*, *per se*, risks connected with bad quality food are not as spectacular and explosive as those of disease outbreak. The deterioration in the quality of the food we consume every day as well as poor nutritional habits set our 'room-temperature reality'. Not that science-based regulation is to blame for this development, but rather that this regulation is not able to properly address such concerns.

The mono-dimensional emphasis of EU food regulation after the BSE crisis concentrates upon potential 'outbreaks', 'scandals' and 'scares', while food-related problems – specifically those connected with nutrition, such as obesity, which in its turn contributes to other devastating health consequences (heart diseases, blood pressure diseases, cancers and strokes) and which in practice affect much wider population – escape any EU regulatory control (Macmaolain 2007). Moreover, the nature of EU regulation on free movement makes it nigh impossible for Member States to intervene independently in order to counteract these negative developments by national means. No significant action can be undertaken against nutritional problems or obesity at any level of European governance. Consequently, European regulation fails to take account of important aspects of the broadest notion of food safety by over-emphasizing exceptional 'crisis' measures while overlooking mundane and everyday threats to the health of consumers. The cause of this imbalance may seem obvious and natural enough – namely the precedence of specta-cular crises over acceptable risks. It is argued here that the regulatory system itself allows for this imbalance to be reproduced indefinitely and contributes to its demonstrated inability to tackle the totality of food safety concerns which it has nonetheless been tasked to fulfil, politically speaking.

Although the BSE experience had already illustrated the limitations of the EU regulatory system for food, it has been subsequent challenges that have put the system to the ultimate test. It is argued here that the current framework fails to respond to the new challenges of the EU market which, since 2004, has become bigger and more diverse, and increasingly contingent on developments at the international level. These two factors push the EU to develop more flexibility to take account of a wider set of concerns in its policy and law making.

Challenges and New Developments: Bigger Europe in a Smaller World

A number of important factors influencing the European market regulation have become much more prominent in recent years but do not seem to be properly reflected in the regulatory development. Developments in global market regulation as well as the last two enlargements deserve special attention here. The last part of this section is devoted to new developments of the EU market regulation, which indicate the first signs in the political and legislative agenda of recognition of a wider set of concerns in the regulatory process.

Europe as a Global Player

Over the last decades, accelerated globalisation has led to an immense increase of international flows of goods and services, capital and people. Interdependence among global trade partners and their institution of regulation for their international trade relations contributes to a gradual reduction of prices, increased consumer choice as well as the removal of unnecessary trade barriers. As a result, the EU economy is entangled not only in a web of multilateral and bilateral trade agreements but also in a complex system of international market rules and standards governing various aspects of trade relations and co-operation. Over the last 50 years the EU has become a key player in the global economic system, accounting for about 30 per cent of global GDP and 20 per cent of global trade flows. This put pressure on the EU to conduct its trade policy in a coherent and responsible manner in order to promote development, economic prosperity and stability in the global village.

When viewed in the global context, the EU legislature has two roles. On the one hand, the EU represents its Member States in negotiating global standards and guarantees their implementation and enforcement. It thus guards the observance of the global rules in the European Union and, in case of non-compliance, acts as their executor. On the other hand, however, the EU should also protect the interests of its Member States before the relevant global institutions and see to it that their particular preferences and reservations are given due consideration. In practice, while the Commission represents the EU and its Member States before the WTO bodies, the Commission does not always take into consideration the positions of all Member States on a given instrument, and takes the responsibility for implementation and enforcement of the commonly agreed WTO rules in all 27 States. In the case of highly sensitive issues, this may prove to be rather uncomfortable. A situation of that kind may, for example, be observed currently in the case of Genetically Modified Organisms (GMOs).

In a brief summary, the conflict between the old (the European Union) and the new (the US, Canada and Argentina) world over the use of genetically modified organisms is an old one. Since 1998 the *de facto* moratorium on the use of GMOs led to suspension of approvals of new GMOs in Europe. In

2003, the USA together with Canada and Argentina initiated a WTO procedure against the EU, arguing that the EU policy on genetically modified organisms created illegal trade restrictions, in particular by: (1) implementation of a *de facto* moratorium on the approval of new biotech crop varieties; (2) failing to approve particular GM crops for which American firms were seeking approval; (3) maintaining a situation where some Member States unilaterally banned the import and marketing of GM crops that had been approved at the EU level. In September 2006 the decision of the WTO Dispute Settlement Panel ruled in favour of the plaintiffs and requested the EU to bring the GMO approval process into line with WTO regulations.[5] The end result of the conflict is still not certain. The reform of the approval procedure did not resolve the main problem: the resistance of the Member States to implement the applicable EU rules. The lifting of unilateral bans introduced by many Member States remains uncertain but is nevertheless necessary to comply with the WTO ruling.

In July 2010 the Commission proposed a new flexible approach to GMOs.[6] The proposed amendment of the Directive 2001/18 on the deliberate release into the environment of GMOs[7] allows individual Member States to restrict or ban GM cultivation on their territory on the basis of a number of social, cultural or ethical considerations specific to that State. The final shape of this reform initiative remains to be seen. There is a risk, however, that in order to get away from the transnational conflict, the EU will allow for new internal conflicts and internal EU regulatory fragmentation. It is also feared that the Member States that may decide to benefit from the new rules and restrict GM cultivation will be exposed to the WTO objections, as the mechanism proposed by the new Commission initiative may well conflict with the global trade regime. Thus the end of the *EC-Biotech* WTO procedure might be the end of its battle but not draw an end to the transatlantic conflict (Bernauer and Aerni 2008; Howse and Horn 2009; Weimer 2010).

One of the reasons for such scepticism is to be found at the roots of the WTO-EU conception: namely, the mismatch of the logics of regulation applied by the two organizations. It is not just about more or less of a precautionary approach to regulation, which was often debated in the context of both the *Biotech* and the *Hormone* Cases, but goes deeper in terms of the philosophy upon which the respective regulatory instruments are based. To take just one example, while the WTO relies strongly on the product/process distinction for governance purposes, Europe mixes the process into the product under the mutual recognition governance. Consequently, the distinction applied by the WTO, which is the basis of various regulatory solutions, is not reflected in implementation by the EU. As the rationale gets lost and blurred it creates a mismatch of regulatory solutions which inevitably lead to conflicts in application. There are many more such problematic differences in regulatory logic between the EU and the WTO (Holmes 2006), and they affect their global co-operation and coexistence. Majone goes as far as to accuse the European Commission of misunderstanding and disregarding international standardisation and neglecting the importance of coexistence (Majone 2005). This affects the Internal Market and its participants as well as the position of the EU as a global player.

The transatlantic GMO conflict also illustrates very well the problematic nature of the dual roles of the European Commission and the EU. On the one hand Europe is becoming larger and more diversified, while the regulation does not seem to respond to those new challenges. On the other hand Europe is a part of a global trade community and has to play by international rules. The first important problem in this context is that Europe stands before the WTO both in its own name and in that of its Member States, and yet it still does not respond to the Member States' concerns. There are situations when the EU implements and obeys the international rules but some States refuse to do so. In such cases the EU will have to use methods of 'disciplining' them. Although it may seem easy in theory, practice shows a lot of resistance and significant incapacity of the European institutions in guaranteeing the Member States' compliance. This is especially important and difficult in the contemporary – enlarged – Europe which, although stronger and more influential, must deal with a whole range of new challenges.

The Challenge of Enlargement

Probably the biggest challenge facing Europe at the turn of the millennium, and probably the most substantial approximation process since the creation of the common market, has been that of European Union enlargements of 2004 and 2007 adding 12 new members. An additional complicating factor is that the new Member States constitute a very specific group, and differ significantly from what the Union had been accustomed to before. Not only do they differ from the old Member States, but are also very different from each other.

It is probably still too soon to draw general long-term conclusions – especially since the accessions of Bulgaria and Romania happened only a couple of years ago – but the statistics illustrate the tremendous impact that these two accessions have had on the volume of food traded in the EU. At the beginning of 2009, the food and drink industry of the EU of 27 States generated a turnover of €913 billion, purchasing and processing 70 per cent of the EU's agricultural production. The sector employed over 4 million people in 2007 and provided for almost 500 million European consumers.[8]

Apart from the volume, the variety of food production methods and consumption traditions as well as the economic value thereof have risen, changing both the size of the market but also its pattern of operation. This new, bigger and more diversified food market may require some modification of regulatory approaches.

Most of the new Member States have only been enjoying their real statehood for slightly more than a decade before their accession. They had been practising democracy and a free market economy for only a short time. The enlargements of 2004 and 2007 were not only politically significant but also economically challenging, requiring far-reaching legal adaptations and presenting both the EU and the 12 new Member States with an enormous adjustment task. This disturbed the fragile balance of relevant EU homogeneity and brought some important and difficult issues back onto the political and legal agenda. No wonder that this (r)evolution presented a strong incentive for legislative reform as well as for a re-evaluation of objectives driving the development of the European legal system.

Significant differences between the old members and the new acceding States – in terms of legislative tradition as much as their stage of economic, social and political development, and their provision of scientific expertise and its role in the decision-making – necessitated a tremendous regulatory effort on both sides. A long and difficult process of legal approximation as well as formal and factual harmonisation of systems of all new Member States forced the adaptation of the EU organisation in order to make it possible to accommodate the variety of the new States without disturbing the operation of the EU legal and institutional architecture, and without obstructing the relatively smooth functioning of the common market. This seems to have affected both the substance and form of European legal instruments, as well as those of the new Member States:

> [G]iven heterogeneous policy legacies in the [M]ember [S]tates as well as the diverse preferences of national governments and other domestic actors, one-size-fits-all solutions are neither politically feasible nor normatively desirable.
>
> (Falkner *et al.* 2005)

The bigger common market is not only bigger in size but also translates to a more diversified market. Thus, regulatory solutions successfully applied in the initial constellation may no longer be able to achieve their aim. They may, in fact, even contradict some of the objectives of the common market. This is because heterogeneity implies slightly different needs and is more difficult to manage. In case of European integration, considering the post-Second World War division of the continent and its consequences for the development of states on both sides of the Iron Curtain, the increase of heterogeneity of the common market after each enlargement (and especially after the 2004 enlargement) was to be expected. Increased mobility and trade was additionally aggravated by the global financial crisis and has made the regulatory exercise in the 'Bigger Europe' seem more difficult and constrained than was ever anticipated (Artis *et al.* 2006; Vaughan-Whitehead 2003). With such increasing diversity, a consistent application of unified solutions will inevitably lead to further dissembedding of market regulation and distance it from the society it is meant to serve.

Under the current EU regulatory framework there are a number of institutionalised instruments to protect diversity. Two important examples can be invoked here. The first legal mechanism aims to protect the origin of specific types of regional products that have received special recognition for their geographical origin, traditional components or production methods, and which could otherwise not be approved by the European authorities. The second mechanism is the procedure under Article 114(5) and (6) TFEU (ex Article 95(5) and (6) TEC), which allows Member States to introduce, after the adoption of a harmonisation measure by the Council or by the Commission, a national provision based on new scientific evidence relating to the protection of the environment or the working environment because of a problem specific to that Member State which has arisen after the adoption of the harmonisation measure. In that case, the Member State shall notify the Commission of the envisaged provisions as well as the grounds for introducing them, and the Commission shall, within a period of six months, approve or reject the national provisions involved, having first verified whether or not they are a means of arbitrary discrimination or a disguised restriction on trade between Member States, and whether or not they constitute an obstacle to the functioning of the Internal Market.

Closer analysis of the application of those instruments, however, has shown that although they ought to be appreciated for the purposes they serve, they have certain important shortcomings which affect their application in the case of enlargement. Firstly and most importantly, case studies have shown that the available legal means of protecting diversity can often turn out to be too inflexible and it happens that in cases where their application would be perfectly in accordance with their overarching aim and regulatory purpose, the stringency of their legal form does not allow their application. Secondly, it is debatable whether the catalogue of values which these legal mechanisms protect should be reviewed. When confronted with their interpretation and gradation by the Commission and the Court, it seems that frequently values of high importance are not treated accordingly, and the 'traditionally' framed market and science dominate the decision making.

There is an evident emergence of problems to which neither the Community method nor the new methods of softer governance are able to respond. Those problems stem both from the increased heterogeneity in the enlarged Union as well as from the internal development of the European Union project, which now encompasses a far broader set of issues than it did at the initial stages of the integration process. In my reading of it, both those phenomena lead to similar concerns that may be described as some form of detachment: detachment of the regulation from the market and the societies it concerns (which, after Polanyi, has been referred to as disembeddedness) and detachment of the economic and the social spheres of regulation, which despite the promises of 'Social Europe' remains an unresolved issue of imbalanced integration.

Even without referring to the last two enlargements, differences in the approach and provision of social welfare policy and social protection were already enormous among the previous 15. In fact, harmonisation of welfare issues, as Scharpf has been pointing out continuously, had already proved too difficult to negotiate. Deliberating on the 'European Social Model' and the facilitation of Member State compromises by means of side-payments and package deals, Scharpf presents an opinion that not all sacrifices can be compensated, and that the difficulty of reaching negotiated agreements increases with the growing heterogeneity of Member States' conditions, interests and preferences. He thus concludes that hope for ever achieving the 'European Social Model' was destroyed already with the very first enlargement, let alone the current situation (Scharpf 2006). The overarching problem of accommodating 'the social' thus remains unresolved. Consequently, it is interesting to examine the possibility of the market regulation embracing 'the social' and by those means reintegrating and re-embedding it to make it more responsive to the contemporary needs of the enlarged EU.

An interesting illustration of this problem was highlighted by Dunn (Dunn 2003) through her idea of the Trojan pigs: referring to Poland's agriculture and food sector. The core problem identified is that international standards do not always create a unified playing field wherever they are applied because they are always embedded in local social, economic and institutional circumstances. Thus externally imposed standards can often function as 'Trojan horses' that enter specific regulatory environments and local sets of conditions, and produce unintended side-effects (Dunn 2003). This is partly because international standards are more than just scientifically developed technical rules for market organisation. They also intervene in particular local societal and economic situations. This is clearly visible in the case of the EU's enlargements.

In implementing European standards, acceding States often act merely as transmitters of supranationally established standards resulting from the balancing of the variety of concerns, risks and interests of the current Member States. Governments face a particular challenge in managing these externally imposed regulations, which in themselves are the result of a balancing of diverse 'foreign' interests and divergent local interests. Holmes, perhaps overstating the situation a little, refers to 'quasi-colonial' arrangements (Holmes 2003). In the longer term, the implementation of such external regulatory regimes may, under the pressure of supranational standardisation and competition, lead to the detachment of local interest from EU and global settings (Bruszt and Stark 2003).

Moreover, such a forced reallocation of resources away from the objectives of the greatest social/public importance in order to live up to regulatory standards developed by an external power, may have detrimental consequences for the local economy as well as society. In the case of the last two enlargements, most of the new Member States had already been overburdened with the high costs of their transition reforms and the restructuring of their economies from being state-planned to becoming free-market systems. Furthermore, the first years of this adjustment process had been very costly both economically and socially.

This was the case particularly with regard to the social costs of transformation. It is little wonder that the extra costs of implementing the externally imposed directives were not welcomed by the acceding States and their societies. Additionally, and unlike in previous enlargements where the EU had focused on post-accession adjustment by new Member States, the enlargements of the twenty-first century carried a requirement that compliance should have been acquired before the planned date of accession, which was then strengthened by strictly applied conditionality (Krause 2006). Moreover, new Member States were required to withdraw their human and economic links with countries to their east, without simultaneously being allowed to replace them or even partially compensate for them by making new human and economic contacts with countries to their west (Holmes 2003).

Thus, the lack of a complete implementation of the Internal Market legislation and some form of selective non-compliance could, in fact, favour the new Member States and their economic and social well-being. For example, Krause, in analysing a Polish case of a lack of implementation – in September 2004 Poland had still failed to implement more than 250 directives, around one-sixth of the total number of Internal Market directives – makes a strong argument in favour of non-compliance (Krause 2006). Her analysis shows that the implementation of directives in new Member States could be seen as undermining their competitive advantage. The problem is that although the implementation of directives carries the promise of an improvement of the overall situation, it is required in all cases irrespective of the costs involved and of the real potential benefit for any implementing country. Thus, in many cases, a failure to comply with certain directives could serve a nation's important long-term goals and so ultimately could benefit it.

Hence, it is important to recognise that the failure of these directives to properly address the situation of the recently admitted States suggests that a 'multifaceted regulatory model' may have been a more useful system, rather than the current reliance on externally imposed mandatory rules. This indicates that, in many areas, current EU policies and legislation require reconsideration, and new Member States might indeed be the catalysts of such reform (Krause 2006). The current EU approach towards acceding states, in Krause's reading of the disparity in the implementation of the CAP provision, in particular, illustrates the extent to which the legislation as well as the accession criteria were driven by the old member states' self-interest and willingness to straightforwardly apply model solutions to the CEE countries. Consequently, this hindered regulatory diversity and innovation, which could have offered solutions to some ongoing European regulatory problems.

In conclusion, it is suggested that some methods for re-embedding European market regulation can be found thorough increasing flexibility and inclusiveness of the current regulatory structure. The existing regulatory framework in the food sector does not leave enough room for flexibility and inclusiveness. The 'social' can only marginally be reflected in the regulatory outcome. In an area such as food regulation, where the social sphere of regulation should logically be among primary concerns, growing detachment is becoming apparent. More than ever before, due to the diversity of local markets, and preferences and concerns in the new Member States, a more appropriate regulatory response is required. One-size-fits-all regulatory solutions could not respond to these concerns and nor, therefore, to their own objectives. The consequences will not only be of a social and political nature, such as protests by farmers or conscious consumers, or diverging opinions among national representatives, but may be more far-reaching such as structured non-compliance.

New Developments in EU Market Regulation

There are ways of lending an ear to a wider set of concerns in the European regulation. Not much of a revolution is needed. Moreover, European regulatory framework can be made more flexible by merely rebalancing the patterns of use of available regulatory instruments, favouring those that allow more flexibility and consideration of local economic and social circumstances. With the growing needs of the bigger European market it is essential to consider reforms in that direction and open up the system instead of maintaining a fictitious assumption that the social sphere of regulation can simply be disregarded by the laws of the market.

This section takes a brief look at developments in the EU political and regulatory environment, which seem to indicate rising criticism for the current regulatory approach. Firstly, although rarely discussed in the literature, there have been voices advocating inclusion of 'concern assessment' within regulatory procedures addressing risks (Dreyer and Renn 2009; Dreyer and Renn 2008; IRGC 2009). According to that view, regulatory decisions should take account of relevant societal concerns through an assessment of 'socio-political ambiguity', considering any divergences in individual perceptions of risk, any amplification made by or due to effects in news and media, as well as social and ethical concerns. It seems that this problem should be given more attention due to the new regulatory challenges, such as diversity and global entanglement mentioned above, but also growing sensitivities of consumers which make the impact-assessment of regulatory decisions even more problematic. A traditional, natural-science-based regulatory approach may be incapable of addressing all the necessary regulatory aspects.

Secondly, policy papers and reform proposals increasingly refer to the need for socio-economic implications to be included in the reformed regulatory process.[9] Moreover, there are areas, where socio-economic concerns have successfully been included in the regulatory scheme. The best example of such development is the new framework for chemicals.[10] Under REACH regulation this is mainly done through two mechanisms. The first channel was created through the establishment of the Committee for Socio-Economic

Analysis,[11] within the European Chemicals Agency, which is responsible for preparing the opinion of the Agency relating to the socio-economic impact of possible legislative action on substances. Thus, it can be seen as platform for non-scientific considerations in the assessment and authorisation process. The second channel gives the applicant the possibility to include socio-economic analysis in the application dossier.[12] This allows for consideration of a wider scope of implications of issuing or rejecting authorisation, already from a very initial stage of the procedure. Under this provision, socio-economic analysis may include the following elements: (1) impact on the industry; (2) impact on associated businesses; (3) impact on consumers, e.g. product prices, availability, choice, quality, as well as effects on health and environment affecting consumers; (4) social implications such as job security or employment; (5) availability and technical and economic feasibility of alternative substances or technologies, as well as social and economic impacts of using them; (6) wider implications on trade, competition and economic development, etc. Instruments allowing inclusion of non-scientific concerns under the REACH regulation prove that there are legal options already developed under EU law. Therefore, claims for enhanced inclusiveness of the food regulation are neither impossible to consider nor technically unfeasible to regulate. There are already legal instruments in force, which might indicate a new wave of developments of EU legislation that responds to the new challenges facing the EU.

The greater variety and complexity of foodstuffs regulation is a result of a wider and more differentiated nature and usage of products, and probably explains further concerns and doubts arising in such a cognitive opening of the system (Luhmann 2004). It is feared that a systematic reform of a large volume of legislation would be required. Such reform, however, does not necessarily have to entail a revolutionary undertaking. It could very well, as previous experience has shown, be taken step by step. Novel food would be the best place to start, not only because the mechanisms used in the chemicals regulation could be imitated more easily due to the nature of the pre-market approval procedures, but also because this is one of the areas where the sensitivities of social and economic implications are the most obvious. With growing demands of the larger and increasingly diversified European market becoming more explicit, one might wonder whether it is the legal framework that does not allow for the introduction of an inclusive approach, or whether it is more a lack of political will. Considering experiences with chemicals regulation, it seems that it is the latter, begging the question of whether this is because the socio-economic implications in the case of food are not considered relevant, or rather because they are considered too problematic.

Conclusions

Regulation of foodstuffs in Europe, whether governing agriculture, market or risks, has created a complex and incoherent system, the various elements of which have been guided by different rationales and methodologies. The attempt to restore coherence and effectiveness of the system after the BSE crisis has managed only to tackle extensively some aspects of risk regulation, in particular with regard to the use of scientific expertise in decision making at the EU level. Many important issues related to trade and socio-economic implications of market regulation, however, were not addressed sufficiently by the reform. Drawbacks of such a fragmented approach to food regulation are now highlighted by important developments that put the system to the test both externally and from within. Participation in the global trade regime combined with new internal developments following the last two enlargements expose the EU regulatory scheme for food to new challenges. This provokes questions as to the need to modify current regulatory approaches and methods, and highlights problems which have not been addressed adequately by EU food governance so far.

This contribution pointed to a number of such problems and concerns. The analysis of those problems juxtaposed with the study of the development over time of regulatory tools used by the EU food governance indicates that some of the new concerns may be addressed with certain tools already at hand. It is therefore suggested that a more inclusive framework should be developed as a means of making minor reforms and of balancing the use of available regulatory mechanisms to better take account of those concerns. Both natural and social science should enable law to find the way out of this gridlock, acknowledge its limitations and redefine its role in risk and market governance in food sectors. If the choice and intensity of regulatory intervention was reconsidered accordingly, it would lead to positive developments and guide the common market closer to the desired embeddedness. Consequently, perhaps it would make it easier to rediscover the middle way between 'unity' and 'diversity'.

In practice, this could mean retaining standardisation in those market sectors where unification is the overarching goal and where there are no sufficiently strong conflicting values and objectives to collide with or override the benefits of unification. On the other hand, in sectors where longstanding practice of common market has allowed market participants to gain enough experience in adhering to the core principles and practices of consumer safety, the stringency of regulation could be relaxed to allow more space for local particularities. The same logic could apply where existing national, regional or professional rules provide a sufficient guarantee of consumer safety and, being tailor-made for the locality in question, they do not disturb other relevant social and economic objectives. In such cases unifying standardisation is, rather, counterproductive, destroying the existing equilibrium and creating a new fictitious state of affairs that does not respond to the local concerns and which thus fails the embeddedness test. In such situations, it may be that instead of imposing supranational unifying regulation, it would be advisable to resort to some form of mutual recognition.

Such a selective approach would open up the system to more flexibility and provide a possibility to respect other legitimate concerns in regulating at least those sectors of the market that would not suffer from a lack of stringent uniform standards. Additionally, there are areas where solutions may be inspired by the REACH example, gradually opening up regulatory decision-making to include a wider set of considerations.

Finally, in these concluding sentences, a further question needs to be posed: is the European Union ready to accept the challenge of socio-economic diversity? This contribution aimed to show that it should be ready, but that in fact it is already slightly delayed: with the increasing diversity that Europe has been facing since 2004, the consistent application of unified solutions will inevitably lead to further disembedding of market regulation, and distance it from the societies it is meant to serve. Consequently, compliance problems will start piling up and there will be more reasons for social discontent. Last but not least, if the trend towards unification continues, Europe risks further loss of diversity, which will be extremely difficult to restore. In the light of the above analysis, room for diversity protection, value inclusiveness and balance of regulatory objectives are already to be found within the European framework. The examples quoted above show additionally that technical possibilities are not so difficult to develop and apply.

It is evident that the mere acknowledgement of the relevance of socio-economic considerations in food regulation will open a Pandora's Box, and initiate a new wave of debate. The fact that the issue is problematic, however, should not mean that it is better to disregard it. Retaining the status quo will probably cause further disembedding of market regulation from the market itself, as well as from its social basis. Delaying the reform until another grand crisis occurs does not sound very helpful either; this time the crisis may be much less spectacular, but much more polygonal.

Notes

1 The retail approach was described and strongly criticized by the 1985 White Paper on the Completion of the Internal Market, COM(85)310 final, and in an accompanying document entitled Completion of the Internal Market: Community Legislation on Foodstuffs, COM(85) 603 final, which will be evaluated below.
2 C 120/78, Rewe-Zentral AG *versus* Bundesmonopolverwaltung für Branntwein, [1979], ECR 649.
3 European Commission, The New Approach to Technical Harmonisation and Standardisation, COM(85) 19 final, 31 January 1985.
4 Regulation 178/2002 of the European Parliament and of the Council of 28 January 2002 laying down the general principles and requirements of food law, establishing the European Food Safety Authority and laying down procedures in matters of food safety OJ 2002 L 31/1.
5 *European Communities – Measures Affecting the Approval and Marketing of Biotech Products*, the Panel Report WT/DS291/R, WT/DS292/R, and WT/DS293/R, issued on 29 September 2006.
6 Proposal for a Regulation of the European Parliament and of the Council amending Directive 2001/18 as regards the possibility for the Member States to restrict or prohibit the cultivation of GMOs in their territory, COM(2010) 375 final, 13 July 2010.

7 Directive 2001/18 of the European Parliament and of the Council on the deliberate release into the environment of genetically modified organisms, OJ 2001 L 106/1.
8 Confederation of the Food and Drink Industries of the EU, *CIAA Annual Report 2008*, 8 April 2009, at p. 4.
9 For example: European Commission Communication. *Preparing for the 'Health Check' of the CAP reform*, COM(2007) 722 final of 20 November 2007; European Commission Communication. *Life Sciences and Biotechnology – A Strategy for Europe*, COM(2002) 27 final of 23 January 2002, especially p. 8 and p. 11–14; European Commission Communication. *Nanosciences and Nanotechnologies: An Action Plan for Europe 2005–2009*, COM(2005) 243 final of 7 June 2005, in particular at p. 8–9.
10 Regulation 1907/2006 concerning the Registration, Evaluation, Authorisation and Restriction of Chemicals (REACH), establishing a European Chemicals Agency, amending Directive 1999/45 and repealing Council Regulation 793/93 and Commission Regulation 1488/94 as well as Council Directive 76/769 and Commission Directives 91/155, 93/67, 93/105 and 2000/21, OJ 2006 L 396.
11 *Supra*, Art. 76.1(d).
12 *Supra*, Art. 62(5).

Bibliography

Artis, M., Banerjee, A. and Marcellino, M. (eds) (2006) *The Central and Eastern European Countries and the European Union*, Cambridge: Cambridge University Press.
Bernauer, T. and Aerni, P. (2008) 'Trade Conflict over Genetically Modified Organisms', in K. Gallagher (ed.), *Handbook on Trade and the Environment*, Cheltenham: Edward Elgar Publishing, 183–93.
Bruszt, L. and Stark, D. (2003) 'Who Counts? Supranational Norms and Societal Needs', *East European Politics and Societies*, 17(1): 74–82.
Chambers, G. (1999) 'The BSE Crisis and the European Parliament, in C. Joerges and E. Vos (eds), *EU Committees: Social Regulation, Law and Politics*, Oxford: Hart Publishing, 95–108.
Dreyer, M. and Renn, O. (2008) 'Some Suggestions for a Structured Approach to Participation In Food Risk Governance with a Special Emphasis on the Assessment Management Interface', in E. Vos (ed.), *European Risk Governance – Its Science, its Inclusiveness and its Effectiveness*, CONNEX Report Series, No. 6, 89–121.
Dreyer, M. and Renn, O. (eds) (2009) *Food Safety Governance: Integrating Science, Precautions and Public Involvement*, Berlin-Heidelberg: Springer-Verlag.
Dunn, E.C. (2003) 'Trojan Pig: Paradoxes of Food Safety Regulation', *Environment and Planning A*, 35: 1493–511.
Falkner, G., Treib, O., Hartlapp, M. and Leiber, S. (2005) *Complying with Europe: EU Harmonisation and Soft Law in the Member States*, Cambridge: Cambridge University Press.
Holmes, P. (2006) 'Trade and "Domestic" Policies: the European Mix', *Journal of European Public Policy*, 13(6): 815–31.
Holmes, S. (2003) 'A European Doppelstaat?', *East European Politics and Societies*, 17(1): 107–18.
Howse, R.L. and Horn, H. (2009) 'European Communities – Measures Affecting the Approval and Marketing of Biotech Products', *World Trade Review*, 8(1): 49–83.

International Risk Governance Council (IRGC), 'Policy Brief: Appropriate Risk Governance Strategies for Nanotechnology Applications in Food and Cosmetics', Geneva 2009, available online: <http://www.irgc.org/IMG/pdf/irgc_nanotechnologies_food_and_cosmetics_policy_brief.pdf> accessed 12 December 2011.

Joerges, C., Schepel, H. and Vos, E. (1999) 'The Law's Problems with the Involvement of Non-governmental Actors in Europe's Legislative Process: The Case of Standardisation under the New Approach', EUI LAW Working Papers, No. 99/9.

Krapohl, S. (2003) 'Risk Regulation in the EU between Interests and Expertise – The Case of BSE', *Journal of European Public Policy*, 10: 189–207.

Krause, K. (2006) 'European Union Directives and Poland: A Case Study', *University of Pennsylvania Journal of International Economic Law*, 27: 155–204.

Lister, C. (1992) *Regulation of Food Products by the European Community*, London: Butterworths.

Little, G. (2001) 'Reports. BSE and the Regulation of Risk', *Modern Law Review*, 64: 730–56.

Luhmann, N. (2004) *Law as a Social System*, Oxford: Oxford University Press (translated by K.A. Ziegert, original title: *Das Recht der Gesellschaft*, Frankfurt am Main, 1993).

Macmaolain, C. (2007) *EU Food Law. Protecting Consumers and Health in a Common Market*, Modern Studies in European Law, Oxford – Portland Oregon: Hart Publishing.

Majone, G. (2005) *Dilemmas of European Integration. The Ambiguities and Pitfalls of Integration by Stealth*, Oxford: Oxford University Press.

Millstone, E. and van Zwanenberg, P. (2001) 'Politics of Expert Advice: Lessons from the Early History of the BSE Saga', *Science and Public Policy*, 28: 99–112.

Neyer, J. (2000) 'The Regulation of Risk and the Power of the People: Lessons from the BSE Crisis', *European Integration online Papers*, 4.

Nikolaidis, K. (2005) 'Globalization with Human Faces: Managed Mutual Recognition and the Free Movement of Professionals', in F. Kostoris Padoa Schioppa (ed.), *The Principle of Mutual Recognition in the European Integration Process*, Basingstoke: Palgrave Macmillan, 129–89.

Pelkmans, J. (1987) 'The New Approach to Technical Harmonisation and Standardisation', *Journal of Common Market Studies*, 25: 249–69.

Polanyi, K. (2001) *The Great Transformation. The Political and Economic Origins of our Time*, Boston: Beacon Press.

Scharpf, F.W. (1999) *Governing in Europe. Effective and Democratic?* Oxford: Oxford University Press.

——(2006) 'The Joint-Decision Trap Revisited', *Journal of Common Market Studies*, 44(4): 845–64.

Schepel, H. (2005) *The Constitution of Private Governance – Product Standards in the Regulation of Integrating Markets*, Oxford: Hart Publishing.

Sullivan, C.D. (1983) *Standards and Standardization. Basic Principles and Application*, New York: Marcel Dekker.

Vaughan-Whitehead, D.C. (2003) *EU Enlargement versus Social Europe? The Uncertain Future of the European Social Model*, Cheltenham: Edward Elgar.

Vincent, K. (2004) '"Mad Cows" and Eurocrats – Community Responses to the BSE Crisis', *European Law Journal*, 10: 499–517.

Vos, E. (2000) 'EU Food Safety Regulation in the Aftermath of the BSE Crisis', *Journal of Consumer Policy*, 23: 227–55.

Weiler, J.H.H. (2005) 'Mutual Recognition, Functional Equivalence and Harmonization in the Evolution of the European Common Market and the WTO', in F. Kostoris Padoa Schioppa (ed.), *The Principle of Mutual Recognition in the European Integration Process*, Basingstoke: Palgrave Macmillan, 25–84.

Weimer, M. (2010) 'What Price Flexibility? – The Recent Commission Proposal to Allow for National 'Opt-Outs' on GMO Cultivation under the Deliberate Release Directive and the Comitology Reform Post-Lisbon', *European Journal of Risk Regulation*, 4: 347–54.

2 EU Risk Governance of 'Cloned Food'

Regulatory Uncertainty between Trade and Non-trade

Maria Weimer

Introduction

The European Union currently faces a public debate on whether food produced from cloned animals and their offspring ('cloned food') might be allowed to circulate within the Internal Market and, if so, under what conditions. As with genetically modified food and feed, or nanotechnology, the EU regulators are confronted with a controversial technology that raises food safety concerns as well as ethical and socio-economic implications. There is no specific legal framework that regulates such products in the EU at a time when food products derived from cloned animals are beginning to be commercialised in countries outside the EU, and might also reach the European market.[1]

This chapter analyses the difficulties of creating a viable legal framework for 'cloned food' in the EU. Designing such frameworks for scientific innovations has become a major challenge for the EU, especially in the area of food safety (Dreyer and Renn 2009). Animal cloning offers an instructive example in this regard. It presents a nebulous combination of challenges to divising effective regulation: scientific uncertainty about risks to human health and the environment; strong international trade interest in the free circulation of 'cloned foods' on the EU market; and other concerns (e.g. ethics) as well as societal resistance and negative consumer preferences that affect the legitimacy of regulatory approaches. This mix of circumstances has hampered the establishment of new legal rules for animal cloning under the EU Novel Foods legal framework.[2] Following the experience of political deadlock and regulatory failure in EU policy on genetically modified organisms (GMOs), animal cloning for food seems to once again test the capability of the EU to create legal frameworks that would ensure the right balance between risk control and accommodation of societal concerns on the one hand, and the promotion of technological innovation and global free trade on the other.

The problems with regulating animal cloning examined here combine a legal perspective with insights from the interdisciplinary research on risk governance, which underlies this volume. Scholars of risk governance have argued in favour of developing more precautionary and inclusive risk governance approaches in response to what has been defined as the problem of 'uncertain risks' (Vos and Everson 2009). 'Uncertain risks' indicate a situation in which the role of science to assess technological risks is limited due to the lack of experience with a particular technology or because it is impossible to characterise a particular hazard. As a consequence, regulating 'uncertain risks' becomes much more politicised demanding a higher responsiveness to other societal or ethical values. The analysis in this chapter shows that regulating animal cloning hinges on dealing with 'uncertain risks'. It also shows that the EU institutions have made an effort to create an inclusive risk governance process, and to respond to various scientific but also ethical concerns over animal cloning. However, the establishment of viable EU legal rules in this area has been afflicted with problems due to the complexity and multi-sectoral nature of animal cloning regulation. Moreover, strong pressure from the international trade market and international trade rules has resulted in persistent institutional disagreement over animal cloning. This shows that legal and political pressure from the EU's external trade partners hampers the willingness and ability of the EU to establish precautionary and responsive risk frameworks.

Animal Cloning and the Pressure of the Global Food Market

As often in legal regulation, the devil lies in the detail. Understanding the current regulatory developments and their implications at an EU level, first requires some contextual information about the technology of animal cloning as well its commercialisation at a global level. 'Animal cloning' in the sense used in the present regulatory discussion is defined as the reproduction of genetically identical 'copies' of an animal through Somatic Cell Nuclear Transfer (SCNT). At present, SCNT is the most commonly used technique for animal cloning, and it allows scientists to create genetic replicas (clones) from adult animals that share the same nuclear gene set as another organism.[3]

The primary commercial use of this technology today and in the near future is in the breeding of farm animals for food production. The benefits of using animal cloning as a breeding technique lie in the potential to produce elite animals to be used in breeding. Thus, the animals to be cloned would be those having traits of interest for farming such as resistance to disease, or characteristics of interest for food production including the quantity of milk, quality of meat, etc.[4] The clones themselves have a low probability of entering the food chain. It is rather their progeny that are used for food production, above all for the production of milk or meat products.[5] 'Progeny of a clone' refers to offspring born from it by sexual reproduction, where at least one of the parents was a clone.[6]

The first animal clone to gain worldwide attention was Dolly the sheep, whose birth was announced in 1997 (Weise 2006). Since then, the SCNT cloning technique has been considerably improved. As previously in the case of biotechnology, the US industry seems to be closest to the commercial use of animal cloning in the mass production of food,[7] and therefore also represents the strongest commercial interest in removing any potential obstacles to the free international trade of animal cloning products. One important step towards free trade – at least in the US market – was the release of a positive risk assessment of food from animal cloning by the US Food and Drug Administration (FDA) in January 2008. The FDA has found that food derived from healthy animal clones and their offspring does not give rise to more risks than food derived from conventionally bred animals (US Food and Drug Administration 2008). Despite this regulatory green light from the FDA, food from cloned animals is still today not being made available to consumers in the US. This is due to the voluntary moratorium on the sale of such products, which was agreed between the US agriculture and food industry and the US Agriculture Department. The voluntary moratorium has been maintained since 2001 until present.[8] It should be noted, however, that the US moratorium does not cover food obtained from the progeny of clones with the consequence that such food can freely be marketed on the US market as well as exported outside of the US.

It follows that food products derived from the offspring of cloned animals (e.g. milk and meat) may already today be imported into the European Union.[9] In addition, the EU also faces imports of other products derived from animals cloned outside the EU, such as embryos or semen from clones that are traded for breeding purposes.[10] This reveals the economic significance of European regulation of animal cloning. Any trade-restrictive EU measures are likely to endanger international imports into the Internal Market, thus threatening new international trade disputes before the World Trade Organisation (WTO).[11]

The Risks of Animal Cloning: Risks, Hazards and Societal Resistance

The regulatory debate on the risks and concerns associated with animal cloning as a new breeding technique in agricultural food production began only recently. Ever since the release of FDA's draft risk assessment on food from cloned animals in December 2006, the European Commission has been collecting information on the use of this technology in order to inform a European policy approach towards animal cloning. In February 2007, Commission President Barroso turned to two different EU expert bodies with a request to produce assessments of the new technology: the European Group on Ethics of science and new technologies (EGE)[12] was asked to assess the ethical implications of cloning animals for food supply; at the same time, the Commission entrusted the European Food Safety Authority (EFSA) with the task of evaluating the impact of the technology on food safety, animal health and welfare, and the environment. Moreover, the Commission's

General Directorate (DG) SANCO, responsible for health and consumer protection, launched a Eurobarometer survey in 2008 on EU consumer attitudes to cloning for food production, the results of which were published in October 2008. The outcome of these consultations was that the concerns identified with regard to food from cloned animals can be summarised in mainly three groups: scientific uncertainty about risks for human health and the environment; manifest hazards for animal health and welfare; and ethical concerns.

Risks to Human Health, the Environment and to Animal Health and Welfare: EFSA Scientific Opinions

On 15 July 2008, EFSA issued its scientific opinion on animal cloning for food supply.[13] By doing so it followed the Commission's request to advise it on food safety, animal health and welfare, and environment implications of live cloned animals obtained through the SCNT technique, including their offspring and products obtained from these animals. However, EFSA limited its evaluation to cattle and pig clones and their progeny, due to the lack of data for the cloning of other species. Before adopting its final opinion, EFSA carried out an online public consultation on its draft opinion and held a stakeholder meeting.[14] Overall, EFSA identified animal health and welfare as the main concerns arising from animal cloning through SCNT due to the fact that the technique often still malfunctions. In contrast, EFSA was unable to identify risks with regard to food safety and the environment.

With regard to animal health and welfare EFSA found that there are significant animal health and welfare issues for surrogate mothers (dams) and clones that can be more frequent and severe than for conventionally bred animals. Surrogate dams suffer from increased pregnancy failure and increased recourse to Caesarean section. Furthermore, the mortality and morbidity rate of clones in the early stage of their development is considerably higher than in sexually reproduced animals. However, clones that survive appear to be normal and healthy. As regards progeny, EFSA found no indication of any abnormal effects with the consequence that the animal health and welfare concerns identified only apply to animal cloning as such, not to the use of clones for conventional breeding through sexual reproduction.

Moreover, with regard to risks to human health EFSA stated that there is no indication based on current knowledge that differences exist in terms of food safety between food products (e.g. meat and milk) from healthy cattle and pig clones and their progeny, compared to those from healthy, conventionally bred animals. However, EFSA also emphasised that there is not enough data at the moment to evaluate whether SCNT has an impact on the immune functions of cloned animals nor, therefore, on their susceptibility to infection. This raises the question of whether – and to what extent – the consumption of meat and milk from cloned animals or their progeny might also lead to an increased human exposure to transmissible agents. This question remains open, and has been referred for further research into the immunological competence of clones.[15]

Finally, regarding implications of animal cloning for the environment, EFSA concluded that no environmental impact is foreseen: that there is no indication that clones or their progeny would pose any new or additional environmental risks compared to conventionally bred animals. However, EFSA has also acknowledged that only limited data is available with regard to the environmental impact.

To conclude, it became apparent that placing 'cloned food' on the EU market at present would entail manifest hazards to the health and welfare of animals involved in the cloning process. Such hazards arise due to the malfunctioning of the SCNT at present. As the negative impact of this technique on the animals involved seems to be proven – increased pregnancy failure for surrogate mothers, higher mortality and morbidity for clones – this type of risks seems to be fairly well defined and measurable.

On the contrary, possible risks to the general environment such as biodiversity and to human health, have been found difficult to define. EFSA, while stating that based on current knowledge it cannot identify any indications of risk to human health or the environment, also acknowledged that the evidence base available to it was small. This is due to the limited number of studies available, the small numbers investigated and the absence of a uniform approach to allow all the issues relevant to the opinion to be addressed.[16] Thus, EFSA explicitly communicated uncertain information on its risk assessment and recommended further research and monitoring on several issues that could not be sufficiently clarified in its scientific assessment, such as the open question of the immunological competence of clones and the potential consequences for humans.[17]

EFSA's findings have, to a certain degree, different implications for foods obtained directly from animal clones and for foods from clone progeny. The scientific uncertainty surrounding the assessment of health and environmental risks was found equally to persist with regard to both types of products. On the contrary, animal health and welfare concerns were only identified with regard to clones and surrogate mothers and, therefore, not for animals obtained through conventional breeding from clone progeny. As will be shown below, this has clear regulatory ramifications for both types of product (see below 4 and 5).

It follows that EFSA was unable to provide definitive answers to all the questions addressed to it by the Commission, which is why in March 2009 the Commission went back to EFSA, asking it to develop its scientific advice further, especially with regard to animal health and welfare of clones. EFSA's second statement was published on 23 June 2009.[18] Whilst including a number of new publications on SCNT, EFSA confirmed overall the findings and recommendations made in its first risk assessment; at the same time it was still unable to remove all the uncertainties. As a consequence, despite its formulation of 'no indication of risks at present' EFSA's risk assessment leaves us with some uncertainties about possible long-term hazards.

Ethical Concerns

The scientifically identified hazards to animal health and welfare together with the uncertainty surrounding risks to human health and the environment are not the only concerns associated with animal cloning. The EU policy debate was further complicated by ethical concerns. Such concerns were voiced mainly by the European Group on Ethics in New Technologies (EGE) and the European citizens interviewed in the Eurobarometer survey.

The EGE is an independent advisory body created by the Commission. Its task is to advise the latter on ethical questions related to sciences and new technologies, either on request of the Commission or on its own initiative. While appointed by the President of the Commission, the EGE members are asked to issue advice independently from any influence.[19] In the case of animal cloning the EGE acted on request from Commission President Barroso, and it adopted its opinion on animal cloning on 16 January 2008. After conducting expert hearings and a round of public comments, as well as organising a round table with representatives from academia, industry, NGOs, civil society and international organisations, the EGE reached the conclusion that doubts existed about the ethical justification for cloning animals for food supply. The group's main conclusion was that:

> considering the current level of suffering and health problems of surrogate dams and animal clones, the EGE has doubts as to whether cloning animals for food supply is ethically justified. Whether this applies also to progeny is open to further scientific research.
>
> The European Group on Ethics in Science and New Technologies,
> Ethical aspects of animal cloning for food supply,
> Opinion No 23 of 16 January 2008, p. 6.

As a consequence, the EGE saw no convincing arguments justifying the cloning of animals for food.[20] It is interesting to note that the EGE only addressed cloning as such. Whether the use of progeny for food production would also raise ethical problems was not addressed. It could be argued, however, that supplying food from the progeny of clones is more or less the direct result of animal cloning, and therefore of the activity which has been found to raise serious ethical doubts. In other words, farm animals are being cloned precisely for the purpose of breeding them with conventional animals, and of using the progeny for food production. It follows that it would not seem too far-fetched to also extend the ethical concerns related to animal suffering/welfare to clone progeny. Yet, the EGE has left this question open.

Following the EGE recommendation, the Commission's DG SANCO launched a Eurobarometer survey to find out more about EU citizens' attitudes towards animal cloning for food production. The results of the survey, published in October 2008,[21] showed that the majority of citizens hold negative views of animal cloning: 84 per cent believe that the long-term effects of animal cloning on nature are unknown; 77 per cent believe that animal cloning might lead to human cloning; 61 per cent think that animal cloning is morally wrong. A majority of interviewees (58 per cent) said that cloning for food production purposes should never be justified while 63 per cent of citizens stated that it was unlikely they would buy meat or milk from cloned animals, even if a trusted source stated that such products were safe to eat. Finally, special labelling for food products from the offspring of clones was favoured by 83 per cent of the interviewees.

Overall, it seems that the issues perceived to be most problematic by the public are the uncertainty of the long-term effects of the technology on nature and the moral justification for using animals for cloning for the purpose of food production. Two moral objections seem particularly pressing: firstly, the 'slippery slope' argument which contests the cloning of animals by comparison with the (im)morality of cloning humans; and secondly, the fear that animals would run the risk of being treated like commodities rather than as living creatures with feelings.

To conclude, the ethical concerns discussed so far are twofold. On the one hand, the ethical justification for animal cloning was questioned in view of animal suffering caused by the malfunctioning of the SCNT at present. In this respect, animal health and welfare represent the cause for ethical concerns. On the other hand, the ethical concerns expressed in the Eurobarometer survey seem to go beyond animal suffering. They raise the general question of whether animal cloning is morally justifiable, especially against the background of human cloning. For public regulation the differentiation between different causes for ethical concerns is relevant in that fundamental ethical concerns over animal cloning are likely to remain relevant even if the animal cloning technique (SCNT) were to be improved in the future, including an improvement to the welfare conditions of the animals involved.

The analysis of the EU debate on animal cloning until present reveals that EU risk regulators have to manage a variety of risks and concerns associated with the use of this new breeding technique for food production. While there are manifest hazards to animal health and welfare, risks to human health and the environment are surrounded by scientific uncertainty due to the lack of data and research on cloned animals. In addition, the management of these risks is complicated because they are deeply intertwined with general ethical concerns about cloning animals for food production, which makes risk regulation in this area politically controversial. In the evaluation of risks and concerns of animal cloning the progeny of clones seems to present a special category. Hazards to animal health and welfare do not directly apply to progeny, because they are related to the process of animal cloning as such while any progeny of clones (first generation) is produced by sexual reproduction of clones and conventional animals. The EGE did not address the question of the ethical justification for using progeny in food production. The scientific uncertainty of risks to human health and the environment, however, extends also to clone progeny.

Reform of Novel Foods Legislation: Institutional Disagreement and Legislative Failure

Animal cloning for food production challenges public regulation in Europe for three main reasons: the uncertainty of risks for human health and the environment; salience of social and ethical values at stake; and the factual and legal difficulties of pursuing effective regulatory policies at national and EU level due to an important international trade dimension. Moreover, law making and regulation in this area provokes increased media attention and mobilises a wide range of non-regulatory stakeholders, such as NGOs, academics, lobby groups and so forth. As a consequence, the European institutions are struggling to draft suitable legal rules for the use of animal cloning for food production.

Seeing that the commercialisation of animal cloning is a relatively new topic at present there is no specific EU legislation in place to regulate food from cloned animals and their progeny.[22] In January 2008 the Commission presented a legislative proposal[23] to amend Regulation 258/97 concerning novel foods and novel foods ingredients (Novel Foods Regulation)[24] in order to include food from animal cloning in its scope. The ensuing legislative procedure[25] has become the main institutional arena for the political struggles over animal cloning in the EU. In the following I will briefly present the regulatory framework of the Novel Foods Regulation, its relevance for animal cloning and the unfolding of the legislative procedure. It should be said from the outset that the legislative process to reform the Novel Foods framework has failed because of institutional disagreement. At the beginning of 2011 the standard legislative procedure ended in a deadlock between the Council and the Parliament over the issue of animal cloning.[26] As a consequence, reform of the Novel Foods legislation is being delayed, and the current Novel Foods Regulation remains in force.

The Novel Foods Regulation currently requires prior authorisation for novel foods: they may only be placed on the Internal Market after having undergone a centralised safety assessment by EFSA. The Novel Foods Regulation indeed covers food derived directly from an animal clone. Article 1, second paragraph at indent (e) of the Novel Foods Regulation provides that all food isolated from animals which has not been obtained by traditional breeding and does not have a history of safe food use is to be considered 'novel food' and so requires an additional safety assessment. The new Commission proposal does not change the status of food obtained directly from cloned animals under this provision but merely clarifies it by stating that all foods from animals to which 'a non-traditional breeding technique not used before May 1997',[27] has been applied – such as animal cloning – should fall within the definition of novel foods. Therefore, when preparing the amendment the Commission was clarifying the legislative status quo with regard to animal cloning rather than changing it.

However, the Commission chose not to include products from clone progeny in the future definition of 'novel foods'. Therefore, imported food products derived from clone progeny (such as meat or milk) may legally be placed on the Internal Market, being subject only to the general food safety requirements of the Regulation 178/2002.[28] As noted above (see 2), the status of such food on the common market is much more significant in economic terms, especially for international trade, since foods from progeny are likely to be present in the majority of foods traded or imported in the EU. There being no difference between the progeny of clones (created through sexual reproduction with non-clones) and animals obtained through conventional breeding, the former would not be considered as animals to which 'a non-traditional breeding technique'[29] has been applied. Consequently, under the Commission's legislative proposal, products derived from progeny could still freely circulate on the Internal Market. Also, no labelling or traceability requirements are foreseen for such products. It follows that the initial approach to animal cloning adopted by the Commission was to submit only food directly obtained from cloned animals to a centralised EU prior-authorisation requirement involving a case-by-case safety assessment by the EFSA for every product to be put on the market in the EU: a similar approach as that applied to the regulation of GMOs but without labelling requirements for products from animal cloning.[30]

The European Parliament, in its legislative resolution from the first reading held in March 2009, suggested that foods from cloned animals (both from clones and their progeny) should be excluded entirely from the scope of application of the Novel Foods Regulation. Instead, the EP demanded that the Commission submit a new legislative proposal effectively banning animal cloning from the food supply chain.[31] Indeed, the European Parliament has played a very active role in the policy debate on animal cloning. Already in September 2008, as a reaction to the expert opinions of EFSA and the EGE, the Parliament issued a resolution on animal cloning[32] calling for a ban on every form of commercialisation of the technology, including imports of related products into the EU. The resolution was supported by the vast majority of MEPs.[33] It demanded that the Commission:

> submit proposals prohibiting for food supply purposes (i) the cloning of animals, (ii) the farming of cloned animals and their offspring, (iii) the placing on the market of meat or dairy products derived from cloned animals or their offspring and (iv) the importing of cloned animals, their offspring, semen and embryos from cloned animals or their offspring, and meat or dairy products derived from cloned animals or their offspring, taking into account the recommendations of EFSA and the EGE.
>
> European Parliament Resolution on the Cloning of
> Animals for Food Supply, 3 September 2008

The Parliament seems to base its position towards animal cloning on a combination of diverse concerns identified in the course of the policy debate. It stresses the animal health and welfare concerns and the related ethical problems while also extending the argument to reject the commercialisation of foods from progeny. In addition, the Parliament also mentions environmental concerns and the insufficiency of scientific research on cloned animals to date. Finally, it invokes socio-economic concerns by stating that animal cloning 'poses a serious threat to the image and substance of the European agricultural model, which is based on product quality, environment-friendly principles, and respect for stringent animal welfare conditions'.[34]

The Council adopted its first reading position on the Commission proposal in March 2010.[35] It took a mediating position between the Commission and the Parliament and proposed the inclusion not only of food produced directly from cloned animals but also of any food produced from their progeny under the scope of the Novel Foods Regulation, thereby extending the prior-authorisation requirement to the latter type of products. While acknowledging that the Novel Foods Regulation cannot adequately manage all aspects of cloning and also mandating a Commission report on all aspects of animal cloning for food production followed, if appropriate, by a legislative proposal, the Council agreed to use this instrument to regulate food from clones and progeny in order to avoid a legal vacuum until more specific legislation be adopted.[36]

Seeing the institutional disagreement over the Commission proposal, the EU legislators continued their search for a common solution.[37] However, the ensuing process did not result in a compromise. In its second reading the Parliament essentially maintained its original position demanding that animal cloning be excluded from the scope of the Novel Foods Regulation.[38] In October 2010 the Commission signalled its willingness to compromise by publishing a Report to the European Parliament and the Council on animal cloning for food production.[39] In this report the Commission suggested, among others, a temporary ban of five years on animal cloning for food production in the EU including a ban on imported clones or imported foods obtained directly from clones. With this proposal the Commission responded to the welfare concerns and ethical objections against animal cloning for food in the EU. However, the Commission refused to also ban imports of foods from clone progeny as it saw no justification of such measure against the background of EU's international trade law commitments. This seems to be a consequence of the EGE opinion, which did not address the question of the ethical justification of using food from progeny (see above). With regard to human health concerns and based on the EFSA opinion, the Commission saw no scientific evidence to justify restrictions on food from clones and food from offspring of clones based on human health concerns.[40] As a result, the Commission proposal foresaw no labelling requirements for foods from progeny either.

The Commission's offer to issue a temporary ban on animal cloning presented an attempt to resolve the legislative deadlock with regard to the

Novel Foods Regulation. However, the Council rejected the Parliament's amendments at second reading and a conciliatory committee was established. Notwithstanding the Commission report, the Council and the Parliament still could not reach an agreement, primarily because while the Parliament insisted on a commitment to label all foods from animal cloning including those from clone progeny, the Council opposed such commitment.[41]

The Legal Aspects of Regulating 'Cloned Food' in the EU

Difficulties of Regulating Animal Cloning Under the Novel Foods Framework

The collapse of the legislative process to amend the Novel Foods Regulation may well be considered as a political failure. The participating institutions have stumbled over the issue of animal cloning, delaying the regulatory overhaul of the Novel Foods framework, which is necessary not only for regulating animal cloning. Crucially, the amendment also aimed to tackle several other new challenges and developments in the area of food safety, notably that of nanotechnology. Notwithstanding these negative ramifications, the failure of the legislative process may also have a positive aspect. There are considerable legal and political difficulties of regulating food from cloned animals under the Novel Foods framework, which might make the search for new regulatory solutions in this area worth the while.

The first difficulty relates to legal basis and scope of application of the Novel Foods Regulation. The latter is based on EU competence to ensure the functioning of the Internal Market (Article 114 TFEU, formerly Article 95 EC Treaty) and it has been enacted to harmonize national regulations aiming at the protection of human health. Therefore, the Regulation itself serves the purpose of ensuring food safety of novel foods by establishing a common EU safety assessment for such products.[42] However, the regulation of 'cloned food' raises concerns that transcend food safety per se (see above). The legal basis and objectives of the Novel Foods Regulation seem too narrow to accommodate the protection of the environment, animal health and welfare, and ethical concerns. The Council has explicitly emphasised this when giving reasons for its legislative position towards the Novel Foods amendment.[43] It should be noted that the European Parliament tried to broaden the legislative objectives of the Novel Foods Regulation to include consumer protection, animal welfare and environmental protection. In addition, the Parliament voted for a wording that would further emphasise the importance of the precautionary principle and that would establish priority of human health protection over the functioning of the internal market.[44] Nevertheless, even if the legislative objectives had been broadened, the legal basis would still remain with Article 114 TFEU, namely the functioning of the Internal Market. The Novel Foods Regulation is an instrument of EU harmonisation, meaning that it aims at furthering the Internal Market in the area of novel foods. This aspect seems to be at odds with concerns over animal health and welfare and the ethical problems associated with 'cloned food' which, as confirmed by the latest Commission report on animal cloning, should prompt a halt on the commercialisation of 'cloned food' altogether.[45]

That being said, the second difficulty of regulating animal cloning under the Novel Foods Regulation is the inadequacy of the prior-authorisation approach in this area. Let us assume that a case-by-case safety assessment would be carried out by EFSA in the future for every product derived from animal cloning,[46] and submitted for authorisation in the EU. What could possibly be the outcome of such individual EFSA risk assessments seeing that the same Agency in previous scientific opinions could not identify any differences in terms of food safety between foods from healthy clones compared with those from healthy conventionally bred animals (see above)? Of course, EFSA's opinions on animal cloning so far have indicated scientific uncertainty and lack of data, and it cannot be excluded that risks to food safety might be identified at a later stage. This could, with a reference to the precautionary principle (Article 191 II TFEU, ex Article 174 II EC-Treaty), potentially justify the use of the prior-authorisation approach to animal cloning even in view of the present insufficiency of scientific evidence of potential health risks. Without going into further detail here, it can be stated that the application of the precautionary principle in EU pre-market authorisations of novel foods has been afflicted with many difficulties, one of them being its high politicisation and contingency (Weimer 2010b). The problematic example of EU regulation of GMOs (Lee 2008; Pollack and Shaffer 2009) comes to mind: seeing strong public and political differences surrounding animal cloning at present, the scenario of the de facto moratorium on EU authorisations of biotech products[47] can also be imagined for future authorisations of 'cloned food' under the Novel Foods Regulation. The parallels are obvious: a stringent prior-authorisation procedure with individual case-by-case assessments, in which the scientific experts do not identify the existence of risks to food safety and public health, while the Member States – fuelled by strong public opposition from at home – are reluctant to approve the entry on the market of the contested products. New EU legislation for 'cloned food' should draw lessons from the negative experience made with the prior-authorisation procedure for GMOs.

Last but not least, a prior-authorisation procedure for 'cloned food' would also not be able to address animal health and welfare nor ethical concerns over animal cloning. Should the suffering of animals used for animal cloning be considered as morally wrong, as indicated by the present policy debate, a prior-authorisation would in no way be able to prevent this wrong from happening. With regard to animal health and welfare, therefore, a (provisional) ban on animal cloning and on the marketing of 'cloned food' indeed seems to be the logical regulatory consequence.

Animal Cloning and WTO Law

As with other areas of EU risk regulation,[48] the regulation of animal cloning is complicated by its multi-level governance dimension: the importance of world trade law for the evolution of risk approaches at a national and EU level (De Búrca and Scott 2001; Everson and Vos 2009). This dimension is of particular importance because of the ongoing tensions between the commitment of the WTO to ensure the freedom of trade on the one hand, and the ever-growing importance of domestic regulations as potential non-tariff barriers to trade on the other. These tensions are especially strong when it comes to the regulation of new, risk-entailing technologies employed in food products. Here, domestic regulators need to design legal frameworks to respond to public health and environmental concerns while at the same time facing considerable pressure to comply with WTO trade rules (Everson and Vos 2009; Hudec 1998; Howse and Türck 2001; Gruszczynski 2010). The drafting of new EU legislation in the area of food safety and consumer protection has become very WTO-sensitive, especially since the recent experience in the WTO disputes EC-Hormones and EC-Biotech, in which the EU failed to defend its public health policies on new food technologies before a WTO dispute settlement body. European attempts to create more responsive, precautionary and transparent governance approaches (including, for example, labelling and traceability schemes) to new food technologies, have encountered strong resistance from its trading partners. The EU policy process on animal cloning is one such case in point.

Future EU restrictive measures on animal cloning are likely to be problematic against the background of WTO rules. A detailed legal assessment of possible future EU regulation of 'cloned food' cannot be undertaken here (Weimer 2009). Rather, I will show the pertinence of WTO rules in this area in order to explain the difficulties of drafting restrictive measures on animal cloning in view of these rules. The applicability of certain WTO rules will strongly depend on the scope and nature of the restrictive measures concerned (total ban, prior-authorisation, labelling schemes, etc), and on the justification for such legislative measures.

Two main justifications for restricting food from cloned animals are being considered in the EU. Firstly, a measure could be based on public health grounds taking into consideration the scientific uncertainty of possible risks indicated by EFSA and applying the precautionary principle. A prior-authorisation scheme under the Novel Foods framework as originally proposed by the Commission – and supported by the Council – would be appropriate here as the primary objective of the Novel Foods Regulation is the protection of human health (see above). A more restrictive measure would be a ban on animal cloning including the placing on the market of all products derived from cloned animals as demanded by the European Parliament. Finally, future measures could target only foods from cloned animals (Commission's position) or they could also include food from clone progeny (supported by both the Council and Parliament).

Having said this, EU restrictions based on public health grounds are likely to qualify as so-called SPS measures under the WTO Agreement on Sanitary and Phytosanitary Measures (SPS Agreement). While this Agreement recognises the right of the members to adopt measures aiming to protect human, animal, or plant life or health (see Preamble and Article 2.1, SPS Agreement), it is also considered a particularly constraining regime because of its strict science-based requirements for the upholding of national food safety regulation (see above, e.g. Articles 2.2. and 5.1, SPS Agreement). In no case of dispute settlement involving the application of the SPS Agreement to domestic food regulations so far have the latter been found to fulfil the requirements of the Agreement.[49] More particularly, no recourse to the precautionary principle to defend domestic food safety measures has been successful so far due to the strict interpretation of Article 5.7 of the SPS Agreement, which is commonly viewed as an expression of that principle in WTO law (Scott 2003: 229; Poli 2007). This said, any scientific justification for restricting the market access of 'cloned food' that is deemed sufficient under the SPS Agreement, seems to be lacking at present. The Commission report on animal cloning states: 'There is no scientific evidence which could justify restrictions on food from clones and food from offspring of clones based on human health concerns'.[50] It would be difficult to justify EU restrictive measures based on public health grounds under the SPS Agreement, whether food from either clones and their progeny are concerned.

Secondly, concerning the suffering of animals used for animal cloning, future EU restrictions could be based on animal welfare and ethical grounds, hence the Commission proposal to issue a temporary ban on the placing on the market of food from cloned animals including imports.[51] In that case the General Agreement on Tariffs and Trade (hereinafter 'GATT') would be pertinent.[52] Under the GATT any regulatory measure should comply with the 'National Treatment' principle laid down in Article III.4 thereof,[53] which prohibits less favourable treatment of like products imported as compared to domestic products. Moreover, the obligation to eliminate quantitative restrictions expressed in Article XI of the GATT should also be respected. However, if future restrictions on animal cloning were found to violate the above-mentioned GATT provisions,[54] they might be justified under Article XX of the GATT, on grounds of public morals or protection of animal health for instance. To uphold these defences, the measures must not be applied in a manner that would constitute a means of arbitrary or unjustifiable discrimination, or a disguised restriction on trade. Also, the measures should be necessary to protect the public interests invoked.

Overall, EU restrictions of animal cloning under the GATT would have more chance of success compared to the SPS Agreement, especially because the measures discussed at present would concern domestic and imported products from animal cloning alike. Under the GATT the main test for

internal regulation is the question of whether it leads to a de facto discrimination against imports,[55] and the Members may be exempted from the
National Treatment obligation where their regulation aims at protecting certain fundamental public goods.[56] Under the SPS Agreement, however,
domestic laws and regulations are submitted to a considerably more rigorous
test of their 'rationality' in the sense of them being scientifically 'sound'.[57]
Seeing the lack of conclusive scientific evidence for human health risks at
present, the application of the SPS Agreement would – necessarily – lead to
less deference being granted to EU decision makers.

Finally, were future EU measures to foresee a labelling scheme for food
products derived from animal cloning, such measures may qualify as technical regulations in the sense of the Technical Barriers to Trade Agreement
(hereinafter 'TBT Agreement') (Conrad 2006). The TBT Agreement, as
compared to the GATT and the SPS Agreement, is known for being more
generous in recognising the importance of certain values when weighed
against the negative effects on trade (Conrad 2006). The TBT Agreement
places emphasis on the obligation of its members to ensure that their technical regulations do not create unnecessary obstacles to international trade,
while recognising a broader range of policies that can legitimately be pursued
through domestic technical regulations.[58] Any 'legitimate' policy may be the
basis for a TBT regulation. Furthermore, the TBT Agreement does not
regulate risk assessments or require regulations to be based on science
(Marceau and Trachtman 2002). As compared to the SPS Agreement,
the TBT Agreement would arguably make less rigorous demands for future
EU measures on animal cloning including labelling and traceability schemes.
The basic test for the EU authorities would be to show that their non-
discriminatory measures are necessary to fulfil a legitimate policy. Seeing
the negative consumer attitudes expressed in the Eurobarometer study,
consumer protection, consumer choice and ethical concerns would seem to
be most pertinent policy justifications. After all, 83 per cent of Europeans
interviewed in this study showed a preference for specific labelling of foods
from cloned animals.

From this admittedly non-exhaustive overview of the WTO legal aspects it
becomes very clear that regulating animal cloning in the EU is a complex
and sensitive issue. Future restrictive measures would need to be based
either on strong scientific evidence indicating clear risks to public health –
something that is lacking at present – or they would need to be justified as
non-discriminatory, origin-neutral measures necessary to protect the health
and welfare of animals and/or public morals. This explains why the Commission prefers the option of a temporary ban on the placing on the market
of foods obtained directly from clones. This measure would apply to
domestic and foreign products alike and would be based on conclusive evidence and expert opinions because both the EFSA and the EGE confirmed
the existence of hazards to animal health as well as of related ethical concerns. The time limit for such a measure would allow for the advancement

of the animal cloning technique to be taken into consideration, and for the restriction to be reconsidered at a later stage.

However, the Commission proposal would not extend the ban to the placing on the market of foods from clone progeny because the Commission sees no ethical justification for such a measure that would pass the test of the GATT 'General Exemptions' clause. A policy approach to progeny seems to be the bone of contention in the present debate on animal cloning. Because food from progeny is being obtained through conventional breeding, the arguments of animal suffering and ethical implications seem to be more difficult to maintain. The EGE did not address this issue. At the same time, farm animals are being cloned precisely for the purpose of breeding them with conventional animals and using their progeny for food production. Therefore, the commercial drive for foods from clone progeny seems to cause animal cloning, and thus also animal suffering, in the first place. This explains why the European Parliament has responded by demanding a ban to include progeny products.

Then again, there seems to be an additional practical difficulty of regulating the progeny of clones. The traceability and labelling of foods from progeny over several generations is considered as difficult to implement in practice. Seeing the lack of mandatory traceability systems in third countries it may seem disproportionate or even de facto impossible, to require all imported foods from clone progeny to be labelled as such. The key issue seems to be that among the millions of food products that are being imported into the EU each year,[59] conventional foods and those obtained from clone progeny seem to be indistinguishable in practice.

Risk Governance, Law and Science Between Trade and not Trade Concerns

The ongoing struggle to regulate animal cloning in the EU illustrates well the problems of contemporary risk regulation. Public regulators have to accommodate both scientific uncertainty and increased social complexity surrounding the employment of new technologies while at the same time comply with international trade rules. This section evaluates the EU regulatory developments in animal cloning from the viewpoint of interdisciplinary research on risk governance. It shows that the stronger the pressure to comply with international trade obligations, such as in the case of foods from clone progeny, the more difficult it is for EU regulators to design a precautionary governance framework and respond to the many scientific but also ethical concerns over animal cloning.

Multi-disciplinary research on risk regulation and risk governance observed that the context of risk regulation in Europe and abroad has considerably changed over the past 20 years, triggering a change in the very nature of risk regulation itself (Löfstedt 2008). Firstly, the nature of technological risks to be evaluated in the regulatory process has become more complex, rendering traditional quantitative risk assessment methodology partially outdated (Van Asselt and Vos 2008: 281). The new types of risk are marked by uncertainties about the chance and/or extent of possible damage or their divergent interpretations. The main change in the evaluation of technological risks today, therefore, is the shift in focus from the question of probability of the occurrence of a known hazard to the question of defining a potential hazard and its nature in the first place.[60] The evaluation of risks to human health and the environment possibly arising from animal cloning, seems instructive in this regard.

Secondly, new sociological insights as to the socially constructed nature of the scientific process produced in the field of science and technology studies, have induced a redefinition of the role of scientific experts in regulatory decision making (Wynne and Jasanoff 2005; Wynne 1992: 111). Such research revealed that the results of scientific evaluation do not necessarily reflect objective facts but are also the outcome of a particular socially influenced framing of the scientific work. Thus, science and technology studies have made clear the importance of social values and worldviews in the scientific work of experts. Last but not least, in the European context, a series of regulatory scandals in the food sector during the 1990s have seriously undermined public trust in the regulatory system including the scientists employed by it. Despite the ensuing overhaul of EU food legislation (Dreyer and Renn 2009; Vos and Wendler 2006) the use of scientific innovations, such as animal cloning, in food production continues to be one of the most controversial topics on the policy agenda of public regulators in the EU.

As a response to these problems scholars of risk governance have developed the notion of 'uncertain risks' (Vos and Everson 2009). This notion can be best understood as indicating a paradigm shift in the late-modern societal understanding of risk: from a positive conception or risk as empowering and controllable (Bernstein 1996) to a negative conception which associates risks with catastrophic and uncontrollable hazards while at the same time acknowledging the limits of science in assessing them (Beck 1986; Giddens 1999). As a consequence of this, 'uncertain risks' also imply that modern risk regulation is a political rather than technocratic exercise and therefore, the primary function of risk regulation today is the reconciliation of scientific rationality and social/ethical values (Everson and Vos 2009).

It follows that instead of making recourse to seemingly strong science, policy makers should address uncertain risks by implementing risk approaches based on precaution, public participation, transparency and a higher responsiveness to societal values affected by the employment of new technologies (Klinke *et al.* 2006; Klinke 2009; Dreyer and Renn 2009). Risk policies and risk governance frameworks designed for new technologies should include the view not only of scientists but also of other stakeholders including civil society representatives, the general public, academics and so forth (International Risk Governance Council 2005). An early inclusion of a broad range of actors into policy formulation is considered as helpful to avoiding public rejection of new technologies at a later stage, as well as resulting difficulties of regulatory enforcement of legal frameworks – an instructive negative example in this regard is the EU legal framework for GMOs (Lee 2008).

The EU policy process on animal cloning shows that implementing such precautionary and inclusive risk approaches in practice can be afflicted with problems especially where public regulators are under strong pressure to comply with the legal obligations imposed by the global trade order. Emergent technologies are being employed in the production of globally traded products, such as foods, promising better or cheaper foods for consumers around the world. Yet at the same time, they raise not only health and environmental concerns but often also wider societal concerns related to divergent cultural and ethical values rooted in different societies. However, free trade rules both at the EU level and at the WTO level often undermine the powers of domestic regulators to respond to such non-trade concerns raised within the respective societies. As we have seen in the previous section, the legal obligations of free trade and non-discrimination between domestic and foreign products demand strong scientific justifications for any restrictive measures, thus favouring the 'scientification' of domestic risk policies (Everson and Vos 2009).

Here lies the dilemma of EU's current struggle to regulate 'cloned food'. It has been observed that where regulation aims at responding to uncertain risks it is very difficult to find conclusive scientific evidence justifying such regulation. This is due to science not being able to adequately quantify an uncertain risk – to calculate the probability of a hazard to occur using statistical data – either because there is no statistical data available (that is to say, no experience with a new technology) or because it is impossible to characterise a particular hazard (Steele 2004; Van Asselt *et al.* 2009). A situation of 'uncertain risks' seems to be the case with food from cloned animals at present. While EFSA could not identify any risks to public health or the environment it also confirmed scientific uncertainty by stating that there is not enough data and that there are methodological difficulties with assessing animal cloning. Despite this, the Commission has interpreted the results of EFSA's work to indicate no scientific evidence of risks.[61] This reflects what scholars of risk governance have described as the uncertainty paradox, which is characteristic of many regimes of contemporary risk regulation (Van Asselt *et al.* 2009). It appears when it is generally recognised that science cannot provide decisive evidence on uncertain risks but policy makers are still increasingly resorting to science for a greater degree of certainty and conclusive evidence.

Law's role in regulating risk-entailing products seems to be an ambiguous one. On the one hand legal rules are needed, and indeed are called upon to harness scientific progress in order to adjust it to societal expectations, controlling the risks this progress can entail. On the other hand, legal decision making typically relies on facts: on what is known based on past societal and technological experience. Thus WTO trade rules seem to create a strong tendency to follow a 'sound science' rationale simply because they presume the objective nature of science as a universal rationality. In the absence of clearly established facts, for example with regard to risks of animal cloning, WTO law seems to foster the uncertainty paradox.

This 'trade versus non-trade' dilemma is clearly evidenced in EU regulation of animal cloning so far. The institutional actors involved, especially the European Commission, have made an effort to design the policy process as an inclusive process of risk governance. The Commission has undertaken to gather a broad basis of information and knowledge including different aspects of animal cloning. Apart from EFSA, which represented the scientific community, the Commission engaged further non-scientific experts gathered within the EGE in order to anchor ethical values within the debate. Moreover, the voice of the European public was included via the Eurobarometer survey mirroring a range of different concerns, be they ethical or about the long-term effects of animal cloning on food safety and the environment. Above all, the Eurobarometer clearly attests to the public desire for consumer choice and transparency in this area. It should also be noted that both expert bodies, EFSA and EGE, have also carried out public and stakeholder consultation as part of their respective assessments of animal cloning (see above). Overall, public participation and the inclusion of different stakeholders representing different societal interests and values, seems to have been a vital element of the policy debate so far, the Commission to a large extent being the promoter of this process.

However, these efforts have not yet resulted in viable legal rules. The disagreement over the issue of clone progeny has led to the failure of the legislative amendment of the Novel Foods Regulation. Here lies the key problem of regulating food from animal cloning. Foods from progeny seem to be much more relevant for international trade than foods from cloned animals. This is so, because animal clones are produced mainly for the purpose of breeding, and are not being used for food production directly. The food products intended for international trade, therefore, most likely derive from clone progeny. At the same time, it becomes more difficult to establish evidence of risks or ethical concerns over foods from progeny, and thus to legally justify a different treatment of the latter when compared to conventionally produced food.

The institutional disagreement over progeny and the failure of the Novel Foods reform reveal that where the Union's legal and political problems with its external trade partners become too pronounced, the Union's willingness and ability to establish precautionary and responsive risk frameworks is hampered. So far, only the Parliament has defended restrictive measures for clone progeny. The Commission was in favour of treating food from progeny as any other conventional foods on the market, rejecting labelling requirements. Also, the conciliation process in the Novel Foods legislative process collapsed because of the refusal on the part of the Council to generally label foods from clone progeny. With the exception of the

Parliament, therefore, the other institutions were not willing to respond to the extent of the broader call for consumer protection and transparency, clearly expressed in the results of the Eurobarometer study on consumer attitudes towards animal cloning.[62]

To conclude, although regulating the uncertain and socially contested risks of animal cloning would call for more precautionary risk approaches, to respond to a broad range of societal concerns, the implementation of such approaches especially with regard to the progeny of clones seems to be highly difficult. The EU authorities seem willing to respond to the ethical concerns related to the commercialisation of food obtained directly from animal clones by restricting or banning such products on the EU market. However, the institutional agreement seems much more difficult to establish when it comes to also regulating the commercialisation of progeny products because of the high legal and practical constraints related to the global trade of these products. Moreover, the pressure of WTO rules, especially those of the SPS Agreement, make it more difficult to restrict the commercialisation of 'cloned food' on grounds of human health and the environment despite scientific uncertainty surrounding the risk assessment of these products. As a result, instead of invoking the precautionary principle, the Commission contributes to maintaining the 'uncertainty paradox' by demanding clear evidence where the ability of science to provide such is limited.

Conclusions

The current difficulties of regulating 'cloned food' in the EU have been analysed here, combining a legal perspective with insights from interdisciplinary research on risk governance. Foods from cloned animals and especially foods from progeny are globally traded products, which makes any EU restrictions thereon a potential cause of new international trade disputes in the WTO. At the same time EU risk regulators have to manage a variety of risks and concerns associated with the use of this new breeding technique for food production. While there are manifest hazards to animal health and welfare, risks to human health and the environment are surrounded by scientific uncertainty due to the lack of data and research on cloned animals. In addition, the management of these risks is complicated because they are tightly intertwined with general ethical concerns about cloning animals for food production, which makes risk regulation in this area politically highly contested.

An attempt to address these concerns within the EU legal framework for novel foods failed as a result of institutional deadlock over how to regulate foods from clone progeny. This seems to be the bone of contention between the EU institutions at the moment. While foods from clone progeny are being traded internationally[63] it becomes more difficult to establish evidence of risks or ethical concerns over those products, and thus to legally justify a different treatment of food from clone progeny as compared to conventionally produced food.

The inadequacy of the Novel Foods Regulation to regulate animal cloning is clear. As an instrument of EU harmonisation it aims at furthering the free circulation of 'cloned foods' on the Internal Market. It is not a suitable instrument to address the various scientific and non-scientific concerns associated with animal cloning. In addition, as experience with EU regulation of GMOs shows, a prior-authorisation procedure at EU level, as foreseen under the Novel Foods Regulation, seems inappropriate to effectively regulate highly controversial risk-entailing products.

The pertinence of WTO rules with regard to animal cloning regulation is equally clear as are the difficulties of drafting restrictive measures on animal cloning in view of these rules. To comply with WTO law, future restrictive measures would need either to be based on strong scientific evidence indicating clear risks to public health – something that is lacking at present – or they would need to be justified as non-discriminatory origin-neutral measures necessary to protect the health and welfare of animals and/or public morals. While such justifications seem possible for measures restricting the free circulation of foods from animal clones, restrictions of foods from clone progeny are likely to run against WTO rules.

As a result, the impact of the WTO legal order on current EU decision making is considerable. EU authorities seem willing to respond to the ethical concerns related to the commercialisation of food obtained directly from animal clones by restricting or banning such products on the EU market. However, the institutional agreement seems much more difficult to establish when it comes to also regulating the commercialisation of clone progeny products because of the legal and practical constraints related to the global trade of these products. Moreover, the pressure of WTO rules, especially those of the SPS Agreement, make it more difficult to restrict the commercialisation of 'cloned food' on grounds of human health and the environment despite scientific uncertainty surrounding the risk assessment of these products. This shows that the stronger the pressure to comply with international trade obligations, the more difficult it is for EU regulators to design and implement a precautionary governance framework and to respond to the panoply of scientific and ethical concerns over animal cloning.

Notes

1 See Report from the Commission to the European Parliament and the Council on animal cloning for food production, from 19 October 2010, COM (2010) 585 final, p. 9.
2 See Regulation 257/97 concerning novel foods and novel food ingredients, *OJ* 1997 L 43/1 (hereinafter 'the Novel Foods Regulation').
3 See in more detail The European Group on Ethics in Science and New Technologies 2008, p. 6 (hereinafter EGE opinion).
4 See EGE opinion 2008, p. 12–13.
5 See European Food Safety Authority (2008: 10) (hereinafter EFSA opinion).
6 See EFSA opinion, 7.

7 USA is the country in which most of the companies aiming to use animal cloning for the food industry have been established. See EGE opinion, *supra* note 3, p. 19; note, however, that other third countries, e.g. Brazil, Argentina, China, Japan are also discussing the commercialisation of animal cloning for food production, see Commission Report on animal cloning, *supra* note 1 at point 5.

8 Refers to the situation as of May 2011. See also Reuters 2008.

9 For an indicative timeline for the commercialisation of food from cloned animals see EGE opinion, *supra* note 3, p. 14 with further references; seeing that food from clone progeny is not being labelled in the US it is at the moment difficult to estimate whether or not it already entered the European market.

10 For example, frozen semen from cloned cattle, bulls and pigs is sold for artificial insemination purposes; for more information see EGE opinion, *supra* note 3, p. 14; see also Commission report on animal cloning, *supra* note 1.

11 Well known precedents are EC – Measures Concerning Meat and Meat Products, Report of the Appellate Body, WTO doc. WT/DS26/AB/R and WT/DS48/AB/R, 16 January 1998; and EC – Measures Affecting the Approval and Marketing of Biotech Products, Report of the Panel, WTO doc. WT/DS291, WT/DS292, WT/DS293, 29 September 2006.

12 The EGE is an independent advisory body created by the Commission. Its last mandate lasted from 2005 to 2009. Its task was to advise the Commission on ethical questions related to sciences and new technologies, either on request of the Commission or on its own initiative; the President of the Commission appointed the EGE members, who were asked to advise the Commission independently from any other influence; see Commission Decision of 11 May 2005 on the renewal of the mandate of the European Group on Ethics in Science and New Technologies, OJ 2005 L 127/17.

13 See EFSA opinion, *supra* note 5.

14 This procedure is part of EFSA's transparency and dialogue strategy, for which it has, among others, institutionalised a stakeholder consultative platform including civil society stakeholders such as NGOs and industry representatives. See Vos and Wendler (2006).

15 See EFSA opinion, *supra* note 5, p. 33.

16 See EFSA opinion, *supra* note 5, p. 2.

17 See EFSA opinion, *supra* note 5, p. 33.

18 See EFSA Statement, 'Further advice on the implications of Animal Cloning (SCNT)', 23 July 2009, *The EFSA Journal* (2009) RN 319, p. 1 et seq.

19 See Commission Decision of 11 May 2005 on the renewal of the mandate of the European Group on Ethics and New Technologies, OJ 2005 L 127/17.

20 See EGE opinion, *supra* note 3.

21 See Eurobarometer, European's attitudes towards animal cloning available online: <http://ec.europa.eu/public_opinion/flash/fl_238_en.pdf > (accessed 2 February 2010).

22 There is, however, EU legislation, which could be applied to regulate certain aspects of animal cloning for food supply. Among the most immediate sources of EU law applicable to animal cloning there are, for example, Directive 98/58 on the protection of animals kept for farming purposes; Directives 91/496 and 97/78 on veterinary checks on the import of animals and products. For further overview see 'Legal aspects of research on and use of farm animal cloning within the EU', report from the project Cloning in Public available online: <http://209.85.129.132/search?q=cache:i5MCLsX-KDQJ:www.sl.kvl.dk/cloninginpublic/+Legal+aspects+of+research+on+and+use+of+farm+animal+cloning+within+the+EU+%E2%80%93+A+synthesis&cd=2&hl=it&ct=clnk&gl=it&client=firefox-a> (accessed 27 July 2009).

23 Commission Proposal for a Regulation of the European Parliament and of the Council on novel foods, COM(2007) 872 final of 14 January 2008, available

online: <http://eur-lex.europa.eu/LexUriServ/LexUriServ.do?uri=COM:2007:0872:FIN:EN:PDF> (accessed 3 February 2010).

24 OJ 1997 L 43, 1–6.

25 Ordinary legislative procedure according to Art. 294, TFEU (formerly co-decision).

26 See on the history of the legislative process EP legislative observatory at <http://www.europarl.europa.eu/oeil/file.jsp?id=5583302> (accessed May 2011).

27 See Commission proposal, *supra* note 23, p. 16.

28 Regulation 178/2002 laying down the general principles and requirements of food law, establishing the European Food Safety Authority and laying down procedures in matters of food safety, OJ 2002 L 31/1.

29 See Commission proposal, *supra* note 23, p. 16.

30 On EU regulation of GMOs, see Lee (2008).

31 See European Parliament legislative resolution of 25 March 2009 on the proposal for a regulation of the European Parliament and of the Council on novel foods available online: <http://www.europarl.europa.eu/oeil/file.jsp?id=5583302> (accessed 3 February 2010).

32 See European Parliament resolution on the cloning of animals for food supply, 3 September 2008, available online: <http://www.europarl.europa.eu/sides/getDoc.do?type=TA&reference=P6-TA-2008-0400&language=EN&ring=B6-2008-0373> (accessed 2 February 2010).

33 622 MEPs voted in favour, 32 against and 25 abstained.

34 See European Parliament resolution, *supra* note 31.

35 Position of the Council at first reading with a view to the adoption of a Regulation of the European Parliament and of the Council on novel foods, amending Regulation 1331/2008 and repealing Regulation 258/97 and Commission Regulation 1852/2001, Interinstitutional file 2008/0002 (COD) of 5 March 2010.

36 See Draft Statement of the Council's Reasons, Interinstitutional file 2008/0002 (COD) of 2 March 2010, p. 7.

37 Following the legislative procedure laid down in Art. 294 TFEU.

38 See European Parliament Resolution of 7 July 2010 on the Council position at first reading for adopting a regulation of the European Parliament and of the Council on novel foods, amending Regulation 1331/2008 and Repealing Regulation 258/97 and Commission Regulation 1852/2001, available online: <http://www.europarl.europa.eu/sides/getDoc.do?type=TA&language=EN&reference=P7-TA-2010-0266> (accessed May 2011).

39 See COM(2010) 585 final of 19 October 2010.

40 See p. 10 of the Commission report on animal cloning, *supra* note 1.

41 The Council agreed to labelling only one type of product: fresh beef obtained from clone progeny: see EP legislative observatory online: <http://www.europarl.europa.eu/oeil/resume.jsp?id=5583302&eventId=1147521&backToCaller=NO&language=en> (accessed May 2011).

42 See recital (2) of the preamble of Regulation 258/97.

43 See Council of the European Union, Draft Statement of the Council's reasons from 2 March 2010, Interinstitutional File 2008/0002 (COD) at p. 7

44 See Position of the European Parliament adopted at second reading on 7 July 2010 with a view to the adoption of Regulation (EU) No. ... /2010 of the European Parliament and the Council on novel foods, recital (1).

45 On the difficulty to enact EU legislation in response to the ethical concerns related to animal cloning see Weimer (2010a: 37).

46 The argument applies for both food from clones and clone progeny.

47 Namely between 1998 and 2004, see Pollack and Shaffer (2009: 68).

48 GMOs, nanotechnology.

49 See cases EC – Measures Concerning Meat and Meat Products, Report of the Appellate Body, WTO doc. WT/DS26/AB/R and WT/DS48/AB/R, 16 January

1998; and EC – Measures Affecting the Approval and Marketing of Biotech Products, Report of the Panel, WTO doc. WT/DS291, WT/DS292, WT/DS293, 29 September 2006; Appellate Body Report, Australia – Measures Affecting Importation of Salmon, WT/DS18/AB/R, adopted 6 November 1998; Panel report, Japan-Measures Affecting Agricultural Products, WT/DS76/R, adopted 27 October 1998; Panel report, Japan-Measures Affecting the Importation of Apples, WT/DS245/R, adopted 15 July 2003.

50 See Commission report on animal cloning, *supra* note 1, p. 10.

51 See ibidem.

52 The text of the original Treaty of 1947 is now incorporated in the General Agreement on Tariffs and Trade of 15 April 1994, in force as of 1 January 1995, 55 UNTS 194, 1867 UNTS 187 (hereinafter 'GATT') and into the Agreement establishing the World Trade Organization, 15 April 1994, in force 1 January 1995, 1867 UNTS 154 (hereinafter 'WTO Agreement').

53 However, a border measure only prohibiting the import of 'cloned food' would be dealt with under Article XI of the GATT.

54 Restrictive measures on food from cloned animals would arguably be considered as so-called 'non-product related process and production measures [PPMs]', the admissibility of which under the GATT is being contested. However, since the US-Shrimp/Turtle WTO ruling such measures can, in principle, be justified under Article XX of the GATT. See in more detail Weimer (2009: 291).

55 See Art. III of the GATT.

56 See Art. XX of the GATT.

57 Regulatory measures shall be based on a scientific 'risk assessment', see Article 5.1 of the Agreement on the Application of Sanitary and Phytosanitary Measures, 15 April 1994, in force 1 January 1995, 1867 UNTS 493 (hereinafter 'SPS Agreement'). Note that in all SPS-related cases to date, WTO tribunals considered the requirements of Article 5.1 in each case finding that they are not fulfilled, see Bernasconi-Osterwalder *et al.* (2006: 261).

58 See Preamble, 5th recital, of TBT Agreement; also Marceau and Trachtman (2002: 832).

59 Note that the EU is the major world player in both imports and exports of food. The total sum of both in 2007 was EURO 138 million, quoted from EUROSTAT statistics in focus, 78/2009, External trade available online: <www.eds-destatis.de/de/downloads/sif/sf_09_078.pdf> (accessed 27 May 2011).

60 An example of this is the potential risks associated with the use of nanotechnology as today there is insufficient knowledge about the possible consequences of this technology on human health and the environment, see IRGC (2005).

61 See p. 10 of the Commission report on animal cloning, *supra* note 1.

62 83 per cent were in favour of labelling.

63 See Commission report on animal cloning, *supra* note 1.

Bibliography

Beck, U. (1986) *Risikogesellschaft: auf dem Weg in eine andere Moderne*, Frankfurt am Main: Suhrkamp.

Bernasconi-Osterwalder, N., Magraw, D., Oliva, M.J., Orellana, M. and Tuerk, E. (2006) *Environment and Trade. A Guide to WTO Jurisprudence*, London: Earthscan.

Bernstein, P.L. (1996) *Against the Gods: the Remarkable Story of Risk*, New York: John Wiley & Sons.

Conrad, C.R. (2006) 'PPMs, the EC-Biotech Dispute and Applicability of the SPS Agreement: Are the Panel's Findings Built on Shaky Ground?', *Hebrew University International Law Research Paper* No. 8–06.

De Búrca, G. and Scott, J. (2001) *The EU and the WTO: Legal and Constitutional Issues*, Oxford: Hart Publishing.

Dreyer, M. and Renn, O. (eds) (2009) *Food Safety Governance*, Berlin/Heidelberg: Springer.

European Food Safety Authority (2008) 'Food Safety, Animal Health and Welfare and Environmental Impact of Animals derived from Cloning by Somatic Cell Nucleus Transfer (SCNT) and their offspring and Products Obtained from those Animals', Scientific Opinion from 15 July 2008, *The EFSA Journal*, 767: 1–49.

——(2009) 'Further advice on the implications of Animal Cloning (SCNT)', 23 July 2009, *The EFSA Journal*, RN 319, 1–15.

European Group on Ethics in Science and New Technologies, 'Ethical Aspects of Animal Cloning for Food Supply', Opinion No. 23 of 16 January 2008.

Everson, M. and Vos, E.I.L. (2008) 'European Risk Governance in a Global Context', in E. Vos (ed.), *European Risk Governance: Its Science, its Inclusiveness and its Effectiveness* (Connex Report Series, 6), Mannheim: Connex, 7–36.

Everson, M. and Vos, E. (2009) 'The Scientification of Politics and the Politicisation of Science', in M. Everson and E. Vos (eds), *Uncertain Risks Regulated*, Oxford: Routledge-Cavendish.

Giddens, A. (1999) 'Risk and Responsibility', *The Modern Law Review*, 62(1): 1–10.

Gruszczynski, L. (2010) *Regulating Health and Environmental Risks under WTO Law*, Oxford-New York: Oxford University Press.

Howse, R. and Türck, E. (2001) 'The WTO Impact on Internal Regulations – A Case Study of the Canada-EC Asbestos Dispute', in G. De Búrca and J. Scott (eds), *The EU and the WTO: Legal and Constitutional Issues*, Oxford-Portland: Hart Publishing.

Hudec, R.E. (1998) 'GATT Constraints on National Regulation: Requiem for an "Aims and Effects" Test', *The International Lawyer*, 32: 619–45.

International Risk Governance Council (2005) *White paper on Risk Governance – Towards an Integrative Approach*, Geneva.

Klinke, A. (2009) 'Inclusive Risk Governance through discourse, deliberation and participation', in M. Everson and E. Vos (eds), *Uncertain Risks Regulated*, Abingdon UK/New York: Routledge-Cavendish, 399–415.

Klinke, A., Dreyer, M., Renn, O., Stirling, A. and Van Zwanenberg, P. (2006) 'Precautionary Risk Regulation in European Governance', *Journal of Risk Research*, 9(4): 373–92.

Lee, M. (2008) *EU Regulation of GMOs*, Cheltenham: Edward Elgar.

Löfstedt, R. (2008) *Risk Management in Post-Trust Societies*, London: Earthscan.

Marceau, G. and Trachtman, J.P. (2002) 'The Technical Barriers to Trade Agreement, the Sanitary and Phytosanitary Measures Agreement, and the General Agreement on Tariffs and Trade: A Map of the World Trade Organization Law of Domestic Regulation of Goods', *Journal of World Trade*, 36(5): 811–81.

Poli, S. (2007) 'The Impact of the "Biotech Dispute" on WTO Law and its Challenges for the European Community', *Yearbook of European Law*, 26: 317–53.

Pollack, M.A. and Shaffer, G.C. (2009) *When Cooperation Fails. The International Law and Politics of Genetically Modified Foods*, Oxford: Oxford University Press.

Reuters, 'No end in site for animal cloning moratorium: USDA', 7 April 2008, available online: <http://www.reuters.com/article/scienceNews/idUSTON77972120080407> (accessed 17 May 2011).

Scott, J. (2003) 'European Regulation of GMOs and the WTO', *Columbia Journal of European Law*, 9: 213–40.

Steele, J. (2004) *Risks and Legal Theory*, Oxford: Hart Publishing.

US Food and Drug Administration, 'Animal Cloning: A Risk Assessment', Opinion from 1 August 2008. Available online: <http://www.fda.gov/AnimalVeterinary/SafetyHealth/AnimalCloning/ucm055489.htm> (accessed 3 February 2010).

Van Asselt, M.B.A. and Vos, E. (2008) 'Wrestling with Uncertain Risks: EU Regulation of GMOs and the Uncertainty Paradox', *Journal of Risk Research*, 11(1–2): 281–300.

Van Asselt, M.B.A., Vos, E. and Rooijackers, B. (2009) 'Science, Knowledge and Uncertainty in EU Risk Regulation', in M. Everson and E. Vos (eds), *Uncertain Risks Regulated*, Abingdon, UK: Routledge-Cavendish, 359–88.

Vos, E. and Everson, M. (eds) (2009) *Uncertain Risks Regulated*, Oxon-New York: Routledge-Cavendish.

Vos, E. and Wendler, F. (2006) *Food Safety Regulation in Europe: A Comparative Institutional Analysis*, Antwerpen-Wageningen, Intersentia.

Weimer, M. (2009) 'Regulating Animal Cloning under WTO Law – Policy Choice versus Science', *Finnish Yearbook of International Law*, 20: 291–332.

——(2010a) 'The Regulatory Challenge of Animal Cloning for Food', *European Journal of Risk Regulation*, 1(1): 31–39.

——(2010b) 'Applying Precaution in Community Authorisation of Genetically Modified Organisms – Challenges and Suggestions for Reform', *European Law Journal*, 16(5): 624–57.

Weise, E. (2006) 'Dolly was World's Hello to Cloning's Possibilities', *USA Today*, 4 July 2006 available online: <http://www.usatoday.com/tech/science/genetics/2006-07-04-dolly-anniversary_x.htm> (accessed 3 February 2010).

Wynne, B. (1992) 'Uncertainty and Environmental Learning: Reconceiving Science and Policy in the Preventive Paradigm', *Global Environmental Change*, 2(2): 111–27.

Wynne, B. and Jasanoff, S. (2005) *Designs on Nature: Science and Democracy in Europe and the United States*, Princeton N.J.: Princeton University Press.

3 Risky Politics

A Sociological Analysis of the WTO Panel on Biotechnological Products[1]

Renata C. Motta

Introduction

> The European Communities, throughout this proceeding, has attempted to remove biotechnology from the context of modern agriculture in order to exaggerate risks and scientific uncertainty. Canada, on the other hand, has sought to put biotechnology squarely back into its proper context.
>
> Canada, cited in World Trade Organization (2006: 193)

> The malicious use of terms has distorted the view in which these products are considered and the way in which they should be treated. Particularly, we would appreciate if the European Communities would restrain itself from using concepts like, "cancer", "may induce dramatic unintended changes", "infestation … to cause contamination", among others.
>
> Argentina, quoted in the World Trade Organization (2006: 199)

The above quotes illustrate the international conflict within the World Trade Organization over the definition of risks from GM foods.[2] Canada and Argentina challenged the emphasis of the European Communities, now the European Union, on the existence of risks associated with these products. Together with the United States of America (USA), these countries are the complaining Parties to this trade dispute. In their view, GM products are not an issue of risk. The trade conflict can be recast as a struggle for definitions, or more precisely a conflict as to how to circumscribe the issue of genetically modified agricultural products in between the logics of politics, economy and science.

In the above citations, Canada was contesting the European policy for GMOs and situated the framework of the problem at the frontiers of the *economy* when referring to modern agriculture, while Argentina highlighted the *scientific* discussion of risk. The European position was portrayed as dealing with the risks within the *political* realm of its GMO policy. All sides rely on science to support their arguments, one emphasizing its accomplishments to ensure the safety of biotechnological products and the other focusing on its limits in the face of uncertainty.

This chapter analyses how both sides of the controversy use political, scientific and economic arguments to dispute interpretations of international law when adjudicating the European Union's policy towards GMOs. The theoretical framework is inspired by Luhmann (2005, 2008) in which he builds a sociological theory of risk based on the primacy of social differentiation. This chapter is an attempt to bridge legal studies and social sciences by using conceptual tools from the social sciences in the analysis and explanation of phenomena usually studied under the auspices of law. Luhmann's systems theory is used to examine how the specific logics of the political, economic and scientific subsystems compete for the interpretation of the international law regarding GMOs.

The empirical material is drawn from the World Trade Organization (WTO) Dispute Settlement System, namely, the Panel report *European Communities – Measures affecting the approval and marketing of biotech products* referenced in the accompanying bibliography. This trade dispute was the longest and most complex case to have ever been treated before the WTO at that time and resulted in a very lengthy report of over 1,000 pages with many and long annexes. It drew heavily on the WTO Agreement on the Application of Sanitary and Phytosanitary Measures (SPS Agreement), which takes a strong science-based approach to disciplining the policies of members of that organization. Therefore, distinguished scholars of international trade law and science studies have expected the results of this panel with special concern about how it would settle a balance between science and democracy:

> Judicial interpretation in the *Biotech Products* case will be critical not only for settling the case at hand, but also for helping to resolve a fundamental tension that lies at the core of the SPS Agreement [...]. Using science to enforce harmonization is a more ambitious goal than using it to combat protectionism, but the SPS Agreement text is ambiguous about how the goals of harmonization and non-discrimination should be balanced. [...] Applying rigid concepts of sound science to scrutinize Member States' regulatory decisions, especially in areas of scientific uncertainty and contested politics, raises problems in terms of cultural autonomy and democratic legitimacy
>
> Winickoff, Jasanoff, Busch, Grove-Write and Wynne 2005: 91

The authors advocate that anti-discrimination should prevail as a principle of the WTO judicial review and not an interpretation of what constitutes scientific sufficiency to foster harmonization of sanitary policies. Such an approach is justified because it is more convergent with the international trade regime declared goals. Driesen (2000) had warned against confounding the 'free trade' concept applied to an internationally agreed regime, the main goal of which is to avoid discrimination among partners with a broad interpretation of that concept implying 'trade free of constraints', i.e. the liberalism

slogan of *laissez-faire*. My argument is that the *Biotech products* Case represents a different interpretation of the goals of the international trade regime,
namely, a neoliberal one. Instead of trade free of constraints, the neoliberal
governmentality (Foucault 2004) implies a specific type of State intervention
in the economy: a demand for regulation that enables the flux of goods. In
contrast to the argument of the mentioned authors (Driesen 2000; Winickoff
et al. 2005), I am not concerned whether the WTO judicial review of the
Biotech products Case means a pressure in the direction of an extreme liberalism or a de-regulation bias. Rather, I contend that what is problematic is
that the multilateral arena constrains democratic members into regulating
pressing issues that the political authorities do not feel they have the legitimacy
to decide upon. My thesis is that this means an intrusion of the economic
timing into the political timing.

In order to demonstrate this argument, the present analysis focuses on
one of the measures under charge: the EU moratorium on GM products. As
mentioned above, the countries in the dispute have accused the European
Communities of suspending the approval of biotechnology products: 'It is at
the critical decision-making junctures, or key stages, of the approval procedure that applications were blocked' (WTO 2006: para. 7.442). In its
defence, the EU argued that there was no moratorium but that the lack of
approvals and delays were the result of prudent and responsible actions and
not a 'decision not to decide'. The alleged moratorium illustrates the risky
political situation of having to decide about risks. Instead of accusing the EU
of establishing a regulatory system for GMOs or of prohibiting those
products, the charge was that the EU did not regulate, neither approving nor
rejecting the products.

Therefore, my main argument is that this case illustrates the interference
of the economic logic in the operation of the political subsystem, obliging
political actors to take decisions despite the specific political constraints
to doing so. The question is not whether the EU approves or prohibits
GMOs, nor whether science is ready to be the basis for either type of
decision. Rather the question is how the political timing to decide on
GMOs is constrained by the economic timing to market such products.
The economic logic uses the international trade regime to set limits to the
political timing to make its operation, i.e. take decisions. I will first describe
some of the background to the case and the selection of the case material
that is examined in detail. In the second section I present the arguments in
dispute in two steps: 1) main challenges made by Argentina, Canada and
the USA, and 2) the arguments used by the EU in its defence. Next, I will
analyse the findings of the WTO Panel regarding the claim that the EU
adopted a moratorium on GMOs. In the fourth section, the dispute is
interpreted in terms of competing logics of the economic, political and scientific subsystems. In the conclusion, I will reflect on the question whether
with the completion of the trade dispute before the WTO, the controversy
about GMO is settled.

Background and Research Approach

The official document from the World Trade Organization, collected at the WTO's archive and which is fully accessible online, is the key source used in the analysis of the trade dispute. The Panel Report entitled *European Communities – Measures affecting the approval and marketing of biotech products* is the final result of the trade disputes WT/DS291/R, WT/DS292/R, WT/DS293/R that took place at the WTO. They refer to three Complainants, respectively, the United States, Canada and Argentina, which submitted their complaints individually against the same Respondent, the then European Communities, now the European Union. On May 2003 the three countries started formal consultations with the EU arguing that some of its measures were affecting their agricultural and food exports and were breaching some of the Union's obligations under the WTO Agreements. As they were not satisfied by the results of these negotiations, in August 2003 the three countries requested the WTO Dispute Settlement Body (which consists of all WTO members) to establish a 'panel of experts' under international trade law in order to consider the case.

Given that the measures under challenge were considered to be the same, one single Panel of legal experts was established to analyse the three complaints. The report was issued on September 2006 and adopted by the Dispute Settlement Body in November 2006. The Panel concluded that three types of measures were involved in the dispute: 1) the general *de facto* EC moratorium on the approvals of biotech products; 2) measures affecting the approval of specific products; 3) the EU Member States safeguard measures against the commercialization of products that had been approved at the European level (WTO 2006: para. 7.98). The final rulings stated that the EU and its Member States had acted in violation of their WTO obligations.

The Parties negotiated the timeframe for the implementation of the recommendations by the EU, which was many times extended. In February 2008, the Dispute Settlement Body authorized the USA to retaliate against the EU for not complying with the rulings, whereas Canada and the EU announced that they had reached a mutually satisfactory solution in July 2009 and Argentina and the EU did the same in March 2010.

The sampling procedure is both theoretically informed as well as based on a first analysis of the material and the case. Section IV on the 'Arguments of the Parties' and its subsections E, F, G and H (respectively referring to the first written submissions of the United States, Canada, Argentina and the EC), were chosen because this is where most arguments about biotechnology applying to agriculture and EU policy sit. In the subsequent sections, legal arguments and cross-references abound. In addition, section VII, which contains the 'Findings', and its subsection D, regarding the results of the moratorium, are analysed.

The material was approached by a qualitative method to interpret and to classify the document into qualitative units. These were constructed before and during the examination of the empirical material (Glaser and Strauss 1967). The arguments under dispute regarding each EU measure were classified into two classes of units. The first set concerns the main issues at stake: decision, time and science. Note that logically these are not exclusive concepts[3] and the contents of arguments sometimes overlap, since it is an open dispute between the actors themselves about how to define what is at stake. The second set of qualitative units classifies arguments into political, economic and scientific types (which not surprisingly correlates to the former set, since taking a decision is a political issue and 'time is money').

The Trade Controversy Over the European GMO Policy

This section divides the WTO dispute into two stages: 1) the three main challenges from USA, Canada and Argentina to the European GMOs policy, and 2) the arguments used by the EU to defend itself.

The first claim made by the United States, Canada and Argentina was that the EU had applied a moratorium on genetically modified products. In the European Communities there are approval procedures for the commercialization of biotech products: they cannot enter the market without previous authorization. The procedure has not been challenged by any of the complaining Parties. For instance, Canada considered that this European model is scientifically based because:

> the nature of the risks associated with biotech products varies considerably from plant variety to variety, general assertions about the risks of biotech products, as a class, cannot be made. Each biotech product needs to be evaluated on a case-by-case basis (...)
>
> WTO 2006: para. 4198

The complaining Parties accused the EU of having suspended its approval procedures by adopting a de facto moratorium on GM products: instead of being evaluated on a case-by-case basis, these are, in practice, banned as a group. The request was that the EU operated its procedures as legally established.

Additionally, the claim was that the suspension of the EU approval system was an intentional discrimination to restrict international trade:

> The *moratorium* disproportionately affects non-EC producers (...) given that majority of biotech products are produced in the United States, Argentina, Canada, and China
>
> WTO 2006: para. 4215

The second claim was that the EC had not succeeded in considering the approval applications for specific GM products without 'undue delay', i.e. without a science-based justification for the delay.

The European Communities' legislation sets deadlines for each of the required steps. It is possible to estimate an approximate length of time within which it seems 'reasonable' that the procedures could be completed. The suspension of procedures has resulted in delays that can in no case be justified in light of the periods of time stipulated in the European Communities' legislation, and these delays are not based on sufficient scientific evidence

WTO 2006: para. 4289

The third and last claim concerned the national measures of some EU Member States against the commercialization in their territories of products approved at the regional level. The main challenge against these measures was that they were not based on science and that they contradicted the existing scientific evaluations.

The USA alleged that, although the Member States had given reasons for their measures, these reasons did not constitute scientific risk assessments and therefore did not justify the measures:

not only have the [EU] Member States failed to produce the requisite risk assessments, they have ignored both the initial risk assessments performed by the [EU] Member States where the applications for approval were filed and the opinions submitted by the European Communities' scientific committees in support of those applications, and, later, the opinions submitted in response to the invocation of the safeguard procedures underpinning the national measures

WTO 2006: para. 4243

The EU began its defence by framing GMOs as a new and complex issue that challenged political authorities. Accordingly, the EU was not an exception but part of an international pattern in which all countries were faced with a double-sided decision: how to enjoy the opportunities that the new technology may bring and at the same time cope with the possible risks.

The European Communities does certainly not seek to impose its prudent approach on other countries, who are free to form their own views on the balance of benefits and risks. Similarly, the present WTO challenge should not be used as a means for the complaining Parties to impose their approach on the European Communities or indeed any other countries, especially at a time where countries around the world are still trying to clarify their respective positions on this complex issue

WTO 2006: para. 4331

However, in order to support its position, the EU emphasized only one side of the issue, namely the threats. The EU not only listed a number of possible adverse effects to health and the environment but also argued that the damage could be irreversible. It drew on scientific arguments to justify its claims. For instance, no technique of genetic modification could control the place of gene insertion or its stability, which could lead to non-intended effects. Moreover, even in case of successful insertion, research indicated the possibility of damage to health and to the environment depending on the type of GMO and its intentional use (WTO 2006: para. 4338). The scientific input in these pronouncements aimed at emphasizing the downside of scientific progress: the simultaneous increase of the unknown, since at the same time as new techniques of genetic modification are developed, ignorance is created as to the place in which the genes are located.

With regard to the charge of delays in the procedures, the EU clarified that, due to the speed of developments in science, the social controversies on its territory, and scientific and regulatory debates at the international level, there was a need to revise its legislation concerning GMOs. The new legislation established that the petitions still pending approval should be re-submitted under the new rules, which would partially explain the delays.

Furthermore, the EU argued that all pending petitions were subjected to requests for additional information on the grounds that the data was insufficient to conduct an appropriate risk assessment according to the legislation. The petitioners were considerably delayed in most cases, according to the EU:

> On a level of principle, the European Communities submits that it is legitimate to request additional information necessary for the completion of a risk assessment and/or compliance with certain standards of risk management or risk communication as they have been established by a regulator and as they apply to the given product in question. That principle applies generally to any product that goes through an approval or inspection procedure designed to ensure that this product is safe. It applies *a fortiori* when the product in issue is based on a new technology which is generally untried and untested and which is recognised by the international Community to have characteristics which inherently require prudence and caution
>
> WTO 2006: para. 4365

While recognizing the existence of delays in its approval procedures, the EU denied the existence of a moratorium. Its argument was that if there were delays, there could be no ban since the system was functioning. Consequently, considerations about delays must also be undertaken specifically with regard to each product, since the system works on a case-by-case basis. Furthermore, even if they would have conceded the existence of a moratorium on approvals of biotech products, the EU claimed that it could not have been challenged before the WTO. On the edge of assuming a failure in the operation of the political subsystem, the EU contends, in this case, that the rules of the commercial subsystem would not apply.

The strategy adopted was to frame the moratorium in a language that would fall beyond the scope of the WTO SPS Agreement, by arguing that it is not a 'measure' but a pattern of behaviour:

> In any event, even assuming that on the basis of that 'evidence', and in spite of the actual facts, it could be said that there was in the past a systematic suspension of the approval process, such a pattern or practice would not as such constitute a challengeable measure under the WTO Agreement
>
> WTO 2006: para. 4372

To escape a judicial revision of the moratorium based on the SPS Agreement on grounds of its science based rules, the EU argues that the scope of the risks that its legislation and policy seeks to address goes beyond the risks covered by that Agreement[4] (WTO 2006: paras. 4334; 4355) and includes the protection of the environment and biodiversity. This strategy narrows the interpretation of the scope of the SPS Agreement and assessed it instead under other WTO Agreements.[5]

To understand this strategy, it is important to highlight the role conferred to science in the WTO Agreements, especially regarding health. This theme entered the international trade agenda due to the perception that States could use health policy arguments as a justification to create barriers to the free circulation of products. The question became more pressing in a context of gradually decreasing quantitative restrictions to trade on goods, such as quotas and tariffs.

Therefore the topic of health risks, referred to as sanitary risks, was integrated in WTO rules with the objective to prevent the economic subsystem from being affected by possibly abusive uses of health policies. At the same time, the rules allow WTO Members to pursue political objectives such as health protection. In establishing an Agreement on the Application of Sanitary and Phytosanitary Measures (SPS Agreement), negotiators tried to balance between the objectives of trade liberalization and health protection, by assigning science the role of mediating between the two subsystems, economy and politics. The main obligation of the SPS Agreement is that health protection measures that may affect international trade are based on scientific principles, and that they should not be maintained without sufficient scientific evidence (Article 2.2). By limiting the operation of the political subsystem in such a way, the actors in the trade regime intend that politics does not excessively interfere in the economy.

The discursive strategy of the EU is thus to frame the risks associated with GMOs as 'uncertain'[6] in order to escape judicial review under the WTO SPS Agreement, which has a very detailed definition of risks and very strict provisions regarding the need for a scientific basis for sanitary measures, including for cases of insufficient information to complete a risk assessment (Article 5.7, disciplining the use of provisional measures). Additionally, the

EU refers to other international rules that apply to the issue: the Biosafety Protocol (WTO 2006: para. 4337). This is because the precautionary principle has more resonance in the international environmental regime than in the trade one (Azevedo 2007; Button 2007) and it is a policy approach preferred by the EU to address the GMOs issue. The European Union 'plays currently a leading role in advancing the precautionary principle, which can be interpreted as an explicit attempt to legitimate decision making on, and regulation of, uncertain risks' (Van Asselt, Vos and Rooijackers 2009: 360). The extent to which the precautionary approach can be accommodated under the SPS disciplines is the object of continuous debate among international trade scholars (Button 2004; Prévost 2007) and is one of the main issues at hand in this case.

In calling for the application of the SPS Agreement to the GMOs case, the Claimants wanted a judicial review that would evaluate the scientific basis of the EU policy for GMOs. The EU prefers that their measures be judged not on their scientific basis but according to trade principles, such as non-discrimination among trade partners. As it will become clear in the next section, the interpretation by the Panel of the European moratorium privileged neither the scientific basis nor the non-discrimination principle.

Table 3.1 The arguments of the countries on dispute

	Charges from Complaining Parties	EC Defence
Decision	– The EU applied a moratorium against the GMOs in that it decided not to take decisions. – The non-operation of the approval system discriminates against imported GM products. – The case must be judged according to the WTO rules.	– There is no moratorium but delays on a case-by-case basis. – This is justified because GMOs are a new risk problematic in which science is evolving and regulatory approaches too. – EU defends its autonomy in the face of WTO rules to choose a prudent approach.
Time	The EU delayed on the consideration of the commercial liberalization of these products without a valid justification for such delays.	The delays were justified, due to the need for more information to conduct a risk assessment, due to changes in the regulatory framework as well as the evolving scientific debate in treating pending applications.
Science	Neither the delays at the regional level nor the safeguard measures instituted by the EU Member Countries are scientifically justifiable.	Science has only started to identify the risks of GMOs and long-term effects are largely unknown.

Source: R.C. Motta

The European Moratorium on GMOs: What Can Justify It?

This section presents the findings of the Panel regarding the claim that the EU had applied a *de facto* moratorium. The relevance of analysing how the international trade rules are applied to this specific charge lies on the eminently political character of what is being adjudicated: the decision not to decide about a policy for GMOs. The question that orients this analysis is how the specific logics of the political, economic and scientific subsystems were taken into account by the WTO Panel to interpret the EC moratorium. The findings of the Panel are thus classified into scientific, economic and political arguments.

Noting that no final decisions had been taken during the period under review, the Panel deferred to the evidence of the moratorium. These pieces showed, according to the complaining Parties and the Panel, that the non-approval was not scientifically substantiated, but politically motivated. Referring to a document of November 2000, the Panel considered that it

> does not support the [EU] argument that there was a standstill because of 'requests [by Member States or the Commission] for additional information on complex issues of risk assessment and management'. Rather, it suggests that *the standstill was the result of public concerns and political debate*, which, according to the document, made it difficult to approve applications
>
> WTO 2006: para. 7.524, emphasis added

Further evidence was used to show that there were political motivations leading to the moratorium, such as the following affirmation of the European Commissioner for Health and Consumer Protection. He states that the European legislation on GMOs of 1990 was issued:

> at a time when concern about GMOs was less obvious. The authorisation procedure became obsolete as consumer concerns grew and consequently, Member States have become more and more reluctant to approve the placing on the market of new GMOs
>
> WTO 2006: para. 7.526a

He said this is an action of the Member States in response to public concerns, albeit science:

> [d]espite our scientific advisors having given the green light for growing and marketing GMO plants and foods, our Member States have blocked new authorisations since 1998
>
> WTO 2006: para. 7.526g

Without scientific justification, the concern of the then European Commissioner for Trade and ironically later Director General of the WTO, was to stop any suspicion of protectionist motivation for the alleged default:

the current moratorium is not plucked out of thin air by the Member States for protectionist reasons: it reflects *the fact that food safety is a highly sensitive and political issue for European citizens*

cited in WTO 2006: para. 7.527a, emphasis added

Since there was neither any scientific basis nor commercial intent behind it, the alleged moratorium would have to have been decided politically. The above quotations emphasize that the political authorities responded to the sensibilities of their constituents regarding GMOs. The perspective of consumers seems to have weighed heavily among the political considerations.

It was not enough, however, that there were no protectionist motives in the EU moratorium. Since delays and suspensions are not free of implications for the international trade the Panel ruled that '[t]he absence of approvals is also a trade issue' (idem).

Thus, the framing of the moratorium as a political issue was overruled by the Panel by interpreting it as a trade issue. After analysing various statements and EU sources, the Panel concluded that they highlight the existence of a *de facto* moratorium which prevented the approval of biotechnology products and 'that action was taken by relevant authorities, or deliberately not taken, so as to prevent approvals for a certain period of time' (WTO 2006: para. 7.534).

Given such a diagnosis, many questions arise as to whether there is any legitimate reason for the moratorium and what may be considered as such. I attempt to answer these questions by classifying the arguments under dispute in the Panel's examination under three headings: the scientific, the economic and political.

First, both the prosecution and the defence used scientific arguments. The EU justified the delays repeatedly in the following terms: 'Member States raised legitimate scientific concerns' (WTO 2006: paras. 7.563, 7.574) and that 'any delay which has occurred is entirely legitimate and related to risk assessment and management considerations' (WTO 2006: para. 7.581). Science was its weapon of defence together with the recourse to the term 'risk assessment', as a legitimate tool of a sanitary policy according to trade rules. The prosecution kept invalidating the scientific justification of the EU. Science was a weapon to counter-argue: 'None of the Member States objecting at the Regulatory Committee offered any competing risk assessment or scientific evidence for their objections' (WTO 2006: para. 7.582).

The Panel concluded that, for some product approval histories, there was no record of requests for further information to the applicant (WTO 2006: para. 7.584), while in others it ruled that the EU objections were not based on a 'scientific evaluation or risk assessment' (WTO 2006: para. 7585). In order to issue its interpretation, the Panel entered into the scientific debate, consulting experts to assist it for that purpose (see, for example, WTO 2006: paras. 7.586–88).[7] It considered issues such as the use of marker genes for antibiotic resistance and a risk assessment for non-target organisms as well as concerns about resistance and tolerance. After falling into the trap of playing the role of 'super-expert' (Van Asselt and Vos 2006: 328),[8] the Panel did not rule on the scientific basis of the moratorium since it was not considered to be a SPS measure but the application of the approval procedures.

Therefore, the Panel addressed the procedural obligations of the SPS Agreement contained in Article 8 and Annex C (1)(a), according to which Members have to assure that control, inspection and approval procedures are undertaken and completed without undue delay. Time became the object of dispute between the competing logics of politics and the economy.

The Parties in this controversy tried to justify their claims regarding the EU moratorium, resorting to the logic of the economic subsystem. The European Union argued that there was removal of a significant number of petitions for commercial reasons and changing strategies. One of the reasons was that 'the request was withdrawn by the applicant with the indication that the applicant preferred to no longer be associated with genetically modified products because of the negative response from the market' (WTO 2006: para. 7.1103).

The counter-argument of the complaining countries was that the business reasons for the withdrawal of the petitions were due not to the market response, but to the high legal risks generated by the moratorium and, concomitantly, due to cost and time factors. According to the Panel, however, purely commercial reasons do not detract from the political authority's right to delay their evaluation provided there is justification to do so. What would be considered as a legitimate justification would have to be evaluated according to the WTO rules.

Finally, the Panel examined arguments of the Parties that resorted to the political logics. The plaintiffs' countries brought as evidence of the moratorium a formal declaration made by five EU Member States (known as the Declaration of the Group of Five), at a meeting of the Council of Ministers in June 1999 (WTO 2006: para. 7474), as proof that these States would act as a minority to block the approval process. The Panel found that the statement shows the intention of the countries that sign it to prevent the approval of GMO products in the EU market.

The prosecution argued further that the European Commission could have approved the product in spite of the blockade of the Group of Five, but resigned the use of its powers because it 'considered that it lacked the necessary political support for completing approval procedures by adopting its own draft measures' (WTO 2006: para. 7.489). In the voting held in the European Regulatory Committee during the period under review, the Panel found that no majority was obtained. Although considering it reasonable, from the political point of view, that in these exceptional circumstances the Commission did not establish new voting procedures given the difficulties of obtaining support from the Member States, the Panel stated that this does not relieve the Commission from using its powers to complete the approval processes.

On the one hand, the Panel acceded to the political logic of deciding about risks while on the other it did not accept it in its judicial ruling of that pattern of behaviour. Thus, even if it considered that it might have been politically immature to reach a full decision on the approval procedures of biotech products, the Panel expected the WTO Member to comply with its obligations. The timing of the international market partners receives precedence over the timing of the political constituents. Therefore, the justifications alleged could not cross the political subsystem to maintain their validity in the international trade regime, informed by the logics of the economic subsystem.

Table 3.2 Scientific, economic and political logics competing for the Panel's interpretation

	European Communities	*USA, Canada and Argentina*	*Conclusions of the WTO Panel*
Scientific arguments	There were delays in decision-making due to considerations of risk assessment and risk management.	The EU countries had not presented an alternative risk assessment to support objections to specific product approvals.	It consulted experts and concluded that the EU did not have scientific evidence not to take a decision. However, since there was no decision, there was not a measure to be examined by the Panel.
Economic arguments	Petitions were withdrawn because the manufacturers did not want to be associated with biotechnology products due to the negative response from the market and changed their strategy.	The withdrawal of the petitions were due to higher legal risks generated by the moratorium and, concomitantly, the cost and time factors.	Purely commercial reasons do not deprive the political authority of the right to delay their evaluation, provided there is justification.
Political arguments	GMOs are a new issue of risk and the political authorities are autonomous to decide how to deal with it without interference from trade rules.	There was a political decision not to take decisions which discriminate against exporters of biotech producers.	The suspension was the result of decisions of public concern and political debate and was a measure adopted by the actions and omissions of the Member States and the European Commission. This political decision has trade effects.

Source: R.C. Motta

Risky Politics: Under Pressure to Decide

This section advances the argument that the WTO Panel review of the EC moratorium did not focus on either its scientific basis or its discriminatory character.[9] Rather, I contend that when analysing the trade effects of the political non-decision of the EU, the WTO experts' ruling advanced the trade liberalization regime by interfering in the autonomy of the political subsystem to decide on the timing to regulate. It demanded that political authorities regulate GMOs even if they do not have the political legitimacy to do so.

The Complainants in the *Biotech products* case did not challenge the EU approval system but claimed that it was not operating. The common expectation of the GM product firms and crops exporters is to have the opportunity that their products can be approved (WTO 2006: para. 4209).

The political subsystem presents itself as a control system, with the tendency to act, compelled by its code to make decisions (Luhmann 2008). It transforms external demands about risk problems in two ways. The first is intervention – defining a risk policy by prohibiting a product or imposing risk management measures, such as mandatory labelling and monitoring – and thereby accepting the consequences that this might have, such as hindrance of technological evolution and market competitiveness, in case of prohibition, or eventual health and environmental damage. The second form is by not intervening or liberalizing the product, when the competent authorities consider that there are no adverse effects that need to be regulated. In the latter, the political authority accepts the consequences in case the future proves otherwise and damages do occur. The issue is not seen as one of risk since it is considered that there is no damage and the opportunities can be freely enjoyed.

Nevertheless, the way in which the political subsystem deals with the subject of risk is, in itself, always risky. Approving the product opens a path of opportunity but the opportunities may not materialize or they can be accompanied by serious and irreversible damage to health and the environment. Not to approve it impedes the existence of possible – but uncertain – opportunities, for fear of potential damage that, in turn, may not materialize. The decision to act in terms of banning biotechnology products or liberalizing them may be qualified in the future as inappropriate if the benefits or the damage are not confirmed.

The clear and predictable rules from the approval procedures inform the economic subsystem about the ways in which the political subsystem can export dangers to it: approving, rejecting or imposing some restrictions to its products. However, when political actors – such as the European Commission in the GMOs case – continuously struggle over a problem and do not reach a decision, from the economic perspective they act in an unexpected manner. The complaining Parties claimed that this political operation violates the commercial rules.

As noted by the Panel, the EU could have completed the approval process but decided not to use its powers to do so. The Panel thus judged the moratorium to be a political act, the result of a decision and, therefore, contingent. The moratorium is, thus, a third way of dealing with risk. It is neither the banning nor the imposition of restrictions, nor unrestricted release: it is the postponement of the decision. And being deliberate, it is itself a political decision: a decision not to decide.

The object of the decision, that is, the issue of the risks of GMOs, is left in the background compared to a meta-political issue: a moratorium is a decision (not) to decide on the issue of risk. It is an example of what Beck (1997) defines as 'political reflexivity' or the politicization of politics, as the institutions established to decide on the commercialization of GMOs are recurrently confronted with challenges to their decisions. Political uncertainty arises as a consequence and the rules of the game have to be constantly negotiated. It is not the risk that is under debate but the decisions about it.

The non-decision leaves the economic subsystem standing to face an uncertain and therefore risky future. The economic investment decisions made by firms in research and product development depend on the prospects for marketing. The uncertainty about the conduct of political authorities responsible for the approval of GMOs is a danger to the economic actors that depend on the normal operation of the subsystem, that is, approving or rejecting products, but always by taking a decision.

This dependency of economic actors on political decisions puts the time factor into evidence: thus the charge of 'delay' to decide. The actors in this international trade forum demand that the actors from the political subsystem justify their lateness. It is important to note that the economic time is different from the political one. The economic actors are in a hurry to commercialize their products in order to obtain returns from investments made, especially in science and development. There is a high degree of coupling between the economic and scientific subsystems: the former rapidly incorporates what is produced at the latter, adding the value of new knowledge to new products. The speed at which the economy adopts new technology poses a problem for the political subsystem, exporting dangers to it.

The societal expectations of safety are addressed to the State. But this is not unproblematic for politics: Krucken (1997) calls this process 'transformation of risk'. He affirms that, in situations of high uncertainty and high pressure to decide, risk management becomes a political risk because the correctness of the decision can only be evaluated in the future. The pressure to decide about new technologies is heightened by the time factor. Already delayed, the political subsystem needs to appeal to science, to evaluate the risks arising from new products, which are ready to enter the market excepting only the approval of the political subsystem. As an external demand that does not properly arise from the scientific agenda itself, scientists tend not to be ready to deliver what the political subsystem is requesting, whether that be in terms of knowledge or in terms of the anticipated time-frame. To the extent that risk assessment is an operation of the scientific subsystem to be used in an external context, as a basis for policy decisions, the scientific subsystem does not take the responsibility to advance in this research agenda. In other words, knowledge about any risks of new products is con-structed in response to demands from the political subsystem and has not originated from the own-functioning logic of scientific research.

The EU rapidly appeals to this discrepancy in the time-frames between politics and science, highlighting the limits of science and reiterating the political autonomy not to wait for science and to make use of precaution. On the one side, the EU expresses its concerns about the long term effects of GMOs and points out that it is only recently that research into the safety of these products began. On the other side, these products are already on the market in other countries. Thus, the political authority stays at the cross-roads of distinct times: economic actors marketing products from which the risks are not researched enough by scientific actors. Thus, the timing of economic actors to market GMOs is quicker than the time needed by sci-entists to know their possible effects. The political authority is faced with getting the timing right in order to decide how to regulate the risks of GMOs. According to the EU there could be a number of ways of dealing with this new risk problematic:

'In light of these risks, governments around the world, since the first commercialisation of GMOs in the early nineties, have started to address the question of how to regulate GMOs. Regulatory approaches range from complete bans to 'laissez faire'. Most, however, consist in setting up an approval system specific to GMOs, based on a case-by-case detailed risk assessment. Often such systems are based on a precau-tionary approach, and decisions are sometimes made dependent on considerations other than scientific factors, such as, for instance, socio-economic considerations. Furthermore, approval may be subject to post-market surveillance requirements. Given the constant evolution of the science on GMOs, regulatory approaches are under constant review in many countries'.

(WTO 2006: para. 4339)

However, this dispute takes place at a trade forum. When the EU is accused before the WTO of 'undue delay' in the consideration of biotechnological products for commercial approval, this means that the justification for a 'due delay' can only be accomplished by referring to the international trade rules. As noted above, the Panel showed its comprehension of the difficulties of the European Commission to use its full powers despite dissent but it was not enough to avoid the judicial review of the WTO body. According to it, the EU was obliged to complete the approval procedures. This is a way of reducing the risky dependency of economic actors on the political decisions.

Table 3.3 The logics of the subsystems struggling to settle the WTO dispute

	Subsystems logics	Relationships to other subsystems	The WTO case
Politics	Political authorities are expected to take decisions: they have both the right and the obligation to define GMO risk policy	Actors from the economic subsystem expect the normal operation of politics, especially to decide if whether to approve or prohibit products	The moratorium is a third way, meaning the unexpected operation of the political subsystem: a decision not to decide or a postponement
Economy	Economic actors are in a hurry to commercialize their products and obtain return from investments, especially in research and development.	Since time is money, delays in decision-making to approve products have economic consequences. Economic actors seek to reduce their dependency on politics by tying it to science (in WTO rules).	– Political actors stressed novelty of the issue and the evolving science, and thus reiterated their autonomy not to wait for science or to take recourse to precaution. – The WTO accorded the right to delay decision provided that it is justified in accordance with trade rules.
Science	Science has its own research agenda. This can be influenced by other subsystems in different degrees.	Science is demanded to meet external objectives: generate knowledge to be incorporated into products, soon ready to enter market; at the same time, evaluate effects of this incorporation performing risk assessments to inform political decisions.	WTO legal experts used scientific experts to help them arbitrate between the logics of the economic and the political subsystems

Source: R.C. Motta

Conclusions

The *Biotech products* case illustrates how the international trade regime advanced beyond the aims of combating protectionism disguised as health protection. The analysis of how the WTO Panel interpreted the EU moratorium on GMOs shows that the multilateral body has not pursued a liberal agenda of de-regulating these products but rather that it conforms more to a neoliberal one in which the economy demands that the State regulate albeit with limits to its interference (Motta 2008). The multilateral arena constrained democratic members to regulating pressing issues that the political authorities did not feel they had the legitimacy to decide upon. My thesis is that it means an intrusion of the economic timing into the political timing.

So, to the core question that imposes itself here. Since the EU has lost the trade dispute before the WTO, has the controversy over the risks been settled thus enabling the political authorities to regulate GMOs smoothly? After more than six years since the conclusion of the trade dispute, the EU has failed to implement all the recommendations of the Panel.[10] Moreover, new bans were implemented by Member States of EU-approved GM products.[11] This reinforces Prévost's conclusion that 'The US may have won this battle, but it seems to have lost the war' (2007: 101). It also indicates that political authorities of the EU Member States stand in defence of their right and autonomy to define and to decide about the risks of GMO products, refracting pressures from trade actors. One can assume, therefore, that the economic subsystem (embodied in the WTO Dispute Settlement System) could not, for now, end this conflict over the risks of those products.

Looking to the future, how might the GMO issue become, if not consensual, then at least less controversial? The framework for risk analysis is designed to put the three subsystems (economy, science and politics) in an interactive process by conferring on politics the role of 'risk manager' that should base its decisions on a scientific risk assessment and take into account economic and consumer interests. Nonetheless risk management is still perceived as a technocratic exercise (Vos and Everson 2009: 6) in which State agencies have competence to decide on 'objective' questions rather than normative and political ones. In this model, the space for democratic deliberation is limited, since it is deemed inefficient to solve technical problems.

This takes us to Habermas's (1968) critique on science and technology as a form of ideology. In technocratic models, politics acquires a negative character: instead of the realization of ends, it aims to solve technical problems and dysfunctions of the market. Politics becomes concerned with means, not with ends. By excluding the ends, the technical solutions do not need to be open to public deliberation, which leads to de-politicization. Transforming political questions into technical ones implies a hidden political domination, since the ends of the rational action are not explicitly stated as choices between alternatives and are given beforehand by the established power relations. This technocratic State makes politics unrecognizable as such.

The application of biotechnology to agriculture was a decision made by market actors, which thus have created a demand for State regulation. It appears as a technical problem to be decided upon – food safety of the insertion of that gene, environmental impact on non-target organisms, rules of co-existence, labelling, etc. – instead of a choice between alternatives to be made. However, this model has been contested in the case of GMOs leading to a politicization of the issue. Many final product consumers as well as farmers (seed consumers) remain unconvinced of the opportunities brought by the new technology and want to decide upon its ends. This is still worse in a situation in which 'uncertain risks' are involved (Van Asselt, Vos and Rooijackers 2009: 359).

Rather than limiting politics to a scientific anchor, a possible compromise to be reached in the case of GMOs tends to go in the direction of more politics. Many risk scholars and researchers from science and technology studies are proponents of this solution and much is being studied about public participation in risk regulation (Callon, Lascoumes and Barthe 2001; Kaufman, Perret, Pettricione, Audétat and Joseph 2004; Latour 2004; Vos and Everson 2009). Luhmann (2008) had already noted that the increasing significance of experiments of participation in risk decision-making indicated that the problem is not one of calculation but its social dimension.

This involves bringing consumers to a decision-making position,[12] which goes well beyond the existence of labelling. It implies deciding on the desirable uses of biotechnology. As long as the risk decision and the advantages of the new technology are restricted to a few market actors (the seeds and pesticide industries) and as long as the introduction of agro-biotechnology seem to threaten alternative decisions about the uses of technology in food production, and while this risky decision is observed as such, it is most likely that GMOs will remain controversial.

Notes

1 This chapter is based on the master thesis presented at the University of Brasilia and was presented at the Young Researchers Workshop 'European Integration between Trade and Non Trade', on 16th April 2010 at the Maastricht University. I would like to thank Fernanda A.F. Sobral for her supervision; Sergio Costa, Laurindo Minhoto and Christina Marques, for their comments on the master thesis; Erika Veiga, Laila Mouwad and Rafael Mafra for the pleasure to discuss the GMOs issue, risk analysis and international trade; Manuel Bastias, for his careful revision and helpful comments; Ellen Vos, Marjolein van Asselt and Michelle Everson for their enthusiasm organizing the Workshop; and Esther Verluis and Denise Prévost for the inspiring feedback.
2 This study relates to the agricultural products obtained with the application of modern biotechnology: seeds or grains for human and animal consumption. In this paper the terms transgenic food, GM food, genetically modified organisms (GMOs) and biotechnogical products will be used interchangeably.
3 For these reasons they cannot be considered as 'categories' in the sense of the content analysis method (Mayring 2010). They are qualitative units designed to classify empirical material according to semantic differences.
4 '[...] in particular, in respect of each of the GMOs the steps which have been taken to protect the environment and to conserve biodiversity are reasonable and legitimate, are not necessarily sanitary or phytosanitary in character, and fall in whole or in part outside the scope of the SPS Agreement; to the extent that any steps taken to protect against risks to human, animal or plant life or health in respect of each of the GMOs could be said to be subject to the SPS Agreement, there has been no undue delay or breach of any part of that Agreement on the part of the European Communities or any Member States, and in any event such steps are provisionally justified on the basis of the insufficiency of scientific evidence' (WTO 2006: para. 4334).
5 I would like to thank Denise Prévost for calling my attention to this.

6 Van Asselt, Vos and Rooijackers (2009: 359) define 'uncertain risks' as follows: 'concern situations of suspected, possible hazards, which are usually associated with complex causalities, large-scale, long-term and trans-border processes, and which are generally difficult to control and also transcend human sensory capacities'.

7 The Panel was composed of 3 Members: Christian Häberli, Mohan Kumar, Akio Shimizu (WTO 2006: para. 1.12). All of them are specialists in international trade law, having participated in the negotiations of trade agreements and/or concentrated their research activities on international economic law. Therefore, they have no specific knowledge about GMOs.

8 In a way similar to how the European Court of First Instance acted in the Pfizer case, according to the authors: 'Clearly the Court got stuck in the deadlock of being forced to do what it may not, and is unable to do. First, it is beyond the Court's competence to go into the scientific merits and validity, as it admits itself. Second, it presupposes that the Court is a "super-expert", [that] would be able to resolve scientific disputes. Leav[ing] alone the fact such a super-expert does not exist; it is anyway impossible for the Court to fulfil this role. However by pronouncing [in] itself on the scientific merits of the plausibility discourses brought forward by the various parties, the Court unconsciously pretends to be the "super-expert" it cannot be' (Van Asselt and Vos 2006).

9 It is not much to emphasize that there was no consensus among European Union actors regarding the adequacy of the policy for GMOs and many Members States favoured – and still do – the adoption of agro-biotechnology. Also, inside each Member State there can be variation among groups, favouring or opposing GMOs. Among petitioners for the approval of biotech products in the EU level there are European firms. Thus, European actors were also affected by the EU measures that did not approve biotech products.

10 For updated information see the website of the WTO, available online: <http://www.wto.org/english/tratop_e/dispu_e/cases_e/ds291_e.htm> (accessed 13 November 2011).

11 According to official information from the EU, available online: <ttp://ec.europa.eu/food/food/biotechnology/qanda/d1_en.htm#d> (accessed 30 March 2010).

12 Polls have shown that consumers are more willing to accept the new technology if associated with some benefit. Miller and Kimmel (2002) compared the Eurobarometer from 1996 and 1999 with a study conducted in 1997 and 1998 in the USA. Questions concerned the usefulness, the risks and the moral acceptability of some biotechnological applications and if the person would support biotechnology adoption to make food protein-richer, more durable or alter the taste or to transform agricultural crops more resistant to insect pests. Regarding usefulness, 53 per cent of the Europeans supported it in 1996, falling to 42 per cent in 1999 contrasted to 60 per cent from North Americans in 1997–1998. Circa 60 per cent of Europeans perceived GM food as risky in both periods compared to 54 per cent of North Americans. Moral acceptability and final support, however, differ substantially: Europeans went from 50 to 37 per cent regarding the moral acceptability whereas two thirds of north-Americans had no objections to it; finally, 42 per cent of Europeans in 1996 and only 30 per cent in 1999 supported biotechnogical application in food, in contrast to 57 per cent of US citizens (Miller and Kimmel 2002). The Eurobarometer report of 2005 confirmed the trend that acceptance raises considerably if GM foods are associated with benefits such as foods with less pesticides (51 per cent) or being environmentally friendlier (49 per cent). Interpreting this poll, Gaskell et al. (2006: 3) concluded that 'The European public is not risk-averse about technological innovations that are seen to promise tangible benefits. (...) The lesson for agri-food biotechnology is that unless new crops and products are seen to have consumer benefits, the public will continue to be sceptical'.

Bibliography

Azevedo, M.N.F. (2007) A OMC e a Reforma Agrícola, Brasilia: Funag.

Beck, U., Giddens, A. and Lash, S. (1994) Reflexive Modernization: Politics, Tradition and Aesthetics in the Modern Social Order, Stanford University Press.

Button, C. (2004) The Power to Protect: Trade, Health and Uncertainty in the WTO, Oxford/Portland, Oregon: Hart Publishing.

Callon, M.; Lascoumes, P. and Barthe Y. (2001) Agir dans un monde incertain – Essai sur la démocratie technique, Paris: Seuil.

Driesen, D.M. (2000) 'What is Free Trade? The Real Issue Lurking Behind the Trade and Environment Debate', available online at the Social Science Research Network Electronic Paper Collection: <http://papers.ssrn.com/paper.taf?abstract_id=217412>.

Foucault, M. (2004) Naissance de la biopolitique, Paris: Gallimard/Seoul.

Gaskell, G.; Stares, S.; Allansdottir, A.; Allum, N.; Corchero, C. and Jackson, J. (2006) Europeans and Biotechnology in 2005: Patterns and Trends. Eurobarometer 64.3, A report to the European Commission's Directorate-General for Research.

Glaser, B. and Strauss, A. (1967) The Discovery of Grounded Theory: Strategies for Qualitative Research, Chicago: Aldine Transaction.

Habermas, J. (2006) Técnica e Ciência como "Ideologia", Lisboa: Edições 70.

Kaufmann, A.; Perret, H.; Petriccione B. Bordogna; Audétat, M. and Joseph C. (2004) 'De la gestion à la négociation des risques: apports des procédures participatives d'évaluation des choix technologiques', Revue européenne des sciences sociales, XLII-130: 109–20.

Latour, B. (2004) Políticas da natureza: como fazer ciência na democracia, São Paulo: EDUSC.

Luhmann, N. (2005) 'Risiko und Gefahr', in N. Luhmann (ed.), Soziologische Aufklärung, Wiesbaden: VS Verl. für Sozialwissenschaften, 131–69.

—(2008) Risk: A Sociological Theory, New Brunswick/New Jersey: Transaction Publishers.

Mayring, P. (2010) Qualitative Inhaltsanalyse: Grundlagen und Techniken, 11th ed., Weinheim: Beltz.

Miller, J. and Kimmel, L. (2002) 'The Politics of Biotechnology in the United States and the European Union: A Conceptual and Empirical Analysis', Paper presented at the Annual Meeting of the American Political Science Association, Boston, Massachusetts on 30 August 2002. Available online: <www.allacademic.com> (accessed 15 July 2008).

Motta, R. (2008) 'Biopolítica e Neoliberalismo: A vigilância sanitária no limite da utilidade para o comércio internacional', Revista de Direito Sanitário, 9: 9–32.

Petriccione, B. (2004) 'De l'utilisation de la notion de risque dans le débat public sur les biotechnologies', Revue européenne des sciences sociales, XLII-130: 73–108.

Prévost, D. (2007) 'Opening Pandora's Box: The Panel's Findings in the EC-Biotech Products Dispute', Legal Issues of Economic Integration, 34(1): 67–101.

Van Asselt, M. and Vos, E. (2006) 'The Precautionary Principle and the Uncertainty Paradox', Journal of Risk Research, 9(4): 313–36.

Van Asselt, M., Vos, E. and Rooijackers, B. (2009), 'Science, Knowledge and Uncertainty in EU Risk Regulation', in M. Everson and E. Vos (eds), Uncertain Risks Regulated, Abingdon: Taylor & Francis, 359–88.

Vos, E. and Everson, M. (2009) 'The Scientification of Politics and the Politicisation of Science', in M. Everson and E. Vos (eds), Uncertain Risks Regulated, Abingdon: Taylor & Francis, 1–17.

Winickoff, D., Jasanoff, S., Busch, L., Grove-White, R. and Wynne, B. (2005) 'Adjudicating the GM Food Wars: Science, Risk and Democracy in World Trade Law', *The Yale Journal of International Law*, 30: 81–123.

World Trade Organization/WTO (2006) *European Communities – Measures Affecting the Approval and Marketing of Biotech Products (WT/DS291/R; WT/DS292/R; WT/DS293/R)*. Genebra: Órgão de Solução de Controvérsias da OMC, 29 September 2006.

4 Pre-empting Precaution

GMO Trade Conflicts, Uncertainty Intolerant Risk Assessment and Precaution-based Risk Management

Boryana Ivanova and Marjolein B.A. van Asselt

Introduction

International trade concerns the exchange of goods and services between countries. Numerous trade conflicts are rooted in the question of whether or not particular goods (might) constitute risks to human health and/or the environment. In such instances, risk assessment is intended to take the sting out of the conflict. Risk assessment is usually considered a technical exercise to be performed by experts, and which will result in an objective evaluation of the risks discussed. Risk assessment is usually viewed as an apolitical stage or endeavour. In this paper, we will challenge such assumptions.

Many of the risks at stake in heated trade conflicts are uncertain, and trade conflicts concerning genetically modified organisms (GMOs) are a case in point. In a trade conflict, does it matter how risk assessments deal with uncertainty? Van Asselt and Vos (2008) have observed that risk assessments can be intolerant of uncertainty: instead of identifying and discussing uncertainties, uncertainties are discarded and evaded. Is it a problem when risk assessments are 'uncertainty intolerant'? Does this give rise to political implications? Given the uncertainty, the precautionary principle may be invoked, which can be considered a risk management principle and enables risk managers to defend various precautionary measures ranging from monitoring requirements to labelling or bans, whether temporary or permanent. As the current regulatory arrangements require that risk management strategies should be supported by a risk assessment (see Millstone, forthcoming for a critical discussion of this regulatory model), whether or not uncertainties are identified in risk assessments might impact upon the decision makers' room for manoeuvre.

In the late 1990s, the issue of US exports of GMOs to the European Union (previously referred to as 'European Communities' in the WTO[1]) became a highly contested issue. Different GMOs were authorized for import into the European Union but various countries issued safeguard measures against GMO release onto their markets or use in their environments. Between 1997 and 2000, countries like Italy, Austria, France, Germany, Greece and Luxembourg refused to allow particular GMOs on their territory, although overall consent had been given. The United States claimed that the way the EU dealt with the import of GMOs since 1998 should be characterized as an unofficial *de facto* moratorium that is unsubstantiated by sound science, while the EU claimed it was legitimate that the precautionary principle be invoked by Member States. In May 2003, the US, Canada and Argentina requested consultations at the World Trade Organization (WTO) and, after the consultations failed to resolve the issue, the US requested the establishment of a WTO Panel to rule on its complaint. The report of the Panel in the *European Communities – Measures Affecting the Approval and Marketing of Biotech Products* dispute (hereinafter 'EC–Biotech') was adopted on 21 November 2006.[2] It ruled in favour of the complaining parties.

In this chapter, we discuss the *EC–Biotech* case from the perspective of whether it matters for such trade conflicts how uncertainty is dealt with in the risk assessments, and if so, to what extent. We first introduce relevant parts of the WTO regulatory framework and discuss the nine product-specific safeguard measures (precautionary bans) that were the subject of the case. We then analyse the reasoning of the *EC–Biotech* panel pertaining to uncertainty, risk assessment and precautionary measures.

Background of the *EC–Biotech* Case

The *EC–Biotech* dispute was adjudicated within the framework of the Agreement on the Application of Sanitary and Phytosanitary Measures (the 'SPS Agreement'). The SPS Agreement is considered as 'the central instrument in the global governance for food safety' (Arcuri et al. 2010a: 285). This agreement recognizes the right of WTO Members to impose quarantine measures that may restrict trade in order to protect human health, animal health or plant life. Sanitary and phytosanitary measures, however, should not be used as disguised restrictions to trade.

The purpose of the SPS Agreement is to clarify which factors should be taken into account when imposing a safeguard measure. Measures to insure food safety and animal and plant health should be based on the analysis and assessment of objective and accurate scientific data (see Sampson 2001). Under Articles 5.1 and 5.7 of the SPS Agreement, the level of protection that a particular country wants to achieve may inform a risk assessment and have an impact on determining whether scientific evidence is insufficient (Arcuri et al. 2010b: 183).

Few provisions within the SPS Agreement offer guidance for risk assessment. Article 2 outlines the basic rights and obligations. Paragraph 2 of this Article states that WTO Members should ensure that 'any sanitary or phytosanitary measure is applied only to the extent necessary to protect human, animal or plant life or health' and is 'based on scientific principles and is not maintained without sufficient scientific evidence, except as provided in paragraph 7 of Article 5'.[3]

According to paragraph 1 of Article 5 of the SPS Agreement, which involves risk assessment guidelines and the determination of the appropriate level of sanitary and phytosanitary protection, Members should ensure that 'their sanitary and phytosanitary measures are based on an assessment, as appropriate to the circumstances, of the risks to human, animal or plant life or health'.[4]

The term 'risk assessment' is defined in Annex A(4) of the SPS Agreement. There are two categories of risk assessment – one concerning pests or disease risk and the other concerning food-borne risks. To justify measures applied to address concerns over pests or disease risks, a risk assessment should evaluate the 'likelihood of entry, establishment or spread of a pest or disease' as well as the potential biological and economic consequences associated with it. For measures intended to protect against food-borne risks, the risk assessment should evaluate the 'potential for adverse effects on human or animal health arising from the presence of additives, contaminants, toxins or disease-causing organisms in food, beverages or feedstuffs'.[5]

As mentioned before, the only exception to the requirement of Article 2.2 is described in Article 5.7. The latter states that 'in cases where relevant scientific evidence is insufficient, a Member may provisionally adopt sanitary or phytosanitary measures on the basis of available pertinent information'. In such circumstances, 'Members shall seek to obtain the additional information necessary for more objective assessment of risk and review the sanitary or phytosanitary measure accordingly within a reasonable period of time'.[6] In the *EC–Hormones* case the WTO Appellate Body, the highest authority, stated that 'Article 5.7 reflects the precautionary principle, and that the precautionary principle as such has not been written into the SPS Agreement as a ground for justifying an SPS measure that is otherwise inconsistent with that Agreement'.[7]

In sum, it can be argued that situations of scientific uncertainty, referred to as 'cases where relevant scientific evidence is insufficient' – and hence the possibility of uncertain risks – are in principle recognized within the WTO regulatory framework. Although the precautionary principle is not a universally accepted principle of international law, it is considered a risk management principle in the SPS Agreement and precautionary strategies with regard to uncertain risks are not by default improper restrictions on trade. According to Flett (2010), however, this interpretation is contested and ridden with disputes: controversy often arises here because the term precaution is being used by different people to mean different things.

The EC Member States' Safeguard Measures

Nine safeguard measures were subject of the *EC–Biotech* case. In total the safeguard measures involved twelve product-specific precautionary bans. They were imposed by six Member States, namely Austria, France, Germany, Greece, Italy and Luxembourg.[8] The precautionary bans that were at issue in the *EC–Biotech* dispute are summarized in Table 4.1.

Table 4.1 Overview of countries and product-specific precautionary bans which were subject of the *EC Biotech* case

	Austria	France	Germany	Greece	Italy	Luxembourg
T25 maize	2000				2000	
Bt-176 maize	1997		2000			1997
MON809 maize					2000	
MON810 maize	1999				2000	
MS1/RF1 oilseed rape		1998				
Topas oilseed rape		1998		1998		
BT-11 maize					2000	

Source: Ivanova and Van Asselt

The GMOs concerned were either genetically modified maize or genetically modified oilseed rape. Three biotech companies were involved: the US-based Monsanto, the German Bayer and the Swiss Syngenta (which later merged with Monsanto). Bayer produced the MS1/RF1 and Topas oilseed rapes, while Syngenta was the producer of Bt-11 maize. The other four genetically modified maizes considered in the *EC–Biotech* case were all produced by Monsanto.

It is clear from Table 4.1 above that there is no pattern. None of the countries issuing safeguard measures banned all seven GMOs and none of the GMOs were banned in all the countries involved. Even when a particular GMO was banned in several countries, there are significant differences in terms of the timing of the precautionary ban. Three out of the six countries (Germany, Greece and Luxembourg) banned just one particular GMO and France just two. So the majority of the EU Member States issuing safeguard measures left five or even six out of the seven controversial GMOs untouched.

There is also no pattern visible in the home base of the producers: none of the EU Member States banned all four GMOs produced by the US-based Monsanto, the French bans were even restricted to GMOs produced in another EU Member State (Germany) and the four Italian bans affected two of the three producers. Thus, the claim of the United States, Canada and Argentina, of a European *de facto* moratorium on GMOs with reference to this set of safeguard measures seems, at first glance, rather exaggerated. Nevertheless, the WTO *EC–Biotech* panel ruled in favour of the complaining parties.

The safeguard measures of six EU Member States were assessed by the Panel in the *EC–Biotech* dispute both within the context of Article 5.1 and Article 5.7 of the SPS Agreement. The former Article sets out the requirements for a regular risk assessment while the latter lays down instructions as to how to assess and deal with risks when there is insufficient scientific information.

The EU Member States measures were adopted on the basis of Article 16 of Directive 90/220 (later replaced by Article 23 of Directive 2001/18) and Article 12 of Regulation 258/97. Directive 90/220 'on the deliberate release of GMOs into the environment' allowed a Member State to refuse the release of GMOs, even if overall consent at the EU level was given, provided the country had 'justifiable reasons' to believe that an approved product constituted a risk to human health or the environment. Regulation 258/97 'concerning novel foods and novel food ingredients' set up a similar process for the authorization of novel foods, including food products containing, consisting of or produced from GMOs. This provided the regulatory basis for so-called safeguard measures. The various EU countries therefore framed their refusal to accept an approved GMO in their country as safeguard measures with reference to Directive 90/220 and Regulation 258/97.

EC–Biotech Panel Findings on Consistency With Article 5.1 of the SPS Agreement

The *EC–Biotech* Panel considered the relevant safeguard measures imposed by the six EU Member States to be sanitary and phytosanitary (SPS) measures. As such, they are subject to the provisions of Article 5.1 of the SPS Agreement. It follows that the safeguard measures must be based on an appropriate risk assessment.

The United States asserted that the safeguard measures of the EU Member States were not based on a risk assessment and were, therefore, inconsistent with Article 5.1. In the view of the US, the EU Member States did not present a risk assessment as defined in Annex A(4) of the SPS Agreement. Even though the Member States have expressed concerns about the potential adverse effects of the relevant products, or biotech products in general, the United States complained that these were not based on risk assessments. The risk assessments put forward for the banned products were those conducted by the Member States to which the product applications were originally submitted, and the risk assessment carried out by the European Communities' own scientific committees,[9] which were favourable to the products. The risk assessment of the EU scientific committees did not raise any concerns with respect to human health or the environment. The complaining parties alleged that even though the Member States provided information in support of their measures, this was rejected by those scientific committees, which upheld their initial favourable risk assessments.[10]

The European Communities, on the other hand, argued that the requirement for an SPS measure to be 'based on' a risk assessment does not necessarily mean that this measure must 'conform to' the risk assessment. The same risk assessment may 'reasonably support' more than one possible SPS measure. The EU also asserted that responsible and representative governments may base their decisions and actions on both a mainstream scientific opinion and divergent scientific views. Therefore, the same risk assessment can reasonably support divergent responses by different governments.

The Panel noted that, in respect of the relevant products before their approval, assessments were performed by the competent authorities (the 'lead CA') of the Member State to which the product application was originally submitted and by the relevant EU scientific committee. These assessments were also reviewed by the EU scientific committee when the Member States adopting the measures provided additional information.

The Panel stated that it was common ground among the Parties to the dispute that the assessments conducted by the lead CA and by the scientific committees, constituted 'risk assessments' within the meaning of Annex A(4) and Article 5.1 of the SPS Agreement. Pursuant to the requirements, they evaluated the likelihood of potential adverse effects on human health and/or the environment as well as the associated potential consequences, according to the proposed use of the specific biotech product under consideration. Moreover, the assessments qualified as risk assessment by the *EC–Biotech* Panel were favourable to the biotech products under dispute.

Requirements for Risk Assessment

The *EC–Biotech* Panel addressed the issue of whether the documents that Member States relied on to justify their safeguard measures met the definition of a risk assessment. The Member States justified their safeguard measures against the various biotech products by reference to concerns including the spread of pollen to cultivated surrounding fields, long-term ecological effects, the potential development of antibiotic resistance and allergenicity and toxicity, gene transfer, undesired effects on nontarget organisms, contamination of traditional oilseed rape by genetically modified oilseed rape and adverse effects on biodiversity. The different Member States put forward so-called 'Reasons documents' in order to justify their concerns and hence their ban of the relevant biotech products. These documents were based on a number of studies.

The Panel argued that the studies relied on by Member States did not evaluate the potential for adverse effects on human or animal health arising from the consumption of specific foods containing or consisting of GMOs. Furthermore, neither study assesses the likelihood of risk posed by pest or diseases. The documents which were provided to the Panel did not contain a discussion of risks associated with this specific concern in the context of the product at issue. The Panel therefore concluded that the 'Reasons documents' which the Member States used to justify their safeguard measures, cannot be qualified as risk assessments within the meaning of Annex A(4) and Article 5.1. In sum, only the assessments conducted by the lead Competent Authority and the EU scientific committees were considered risk assessments satisfying Annex A(4) and Article 5.1.

In defence of its measures, the European Union argued that responsible and representative governments may act either on the basis of mainstream scientific opinion or on the basis of a divergent scientific view. The European Union asserted that the safeguard measures were adopted on the basis of new scientific information which presented a view that diverged from the mainstream scientific opinion of the original risk assessment.

As mentioned above, the Panel concluded that the documents put forward by the Member States to justify their safeguard measures are not in themselves risk assessments within the meaning of Annex A(4) and Article 5.1. It stated:

> In the case at hand, we are not aware, and have not been made aware, of any divergent views that would be expressed in the risk assessments of the lead CA and SCP concerning [the measures at issue].[11]

The Panel concluded that there were no divergent opinions expressed in the risk assessments which were conducted by the lead CA and the EU scientific committee. Expressing divergent scientific views would be one way to identify and communicate uncertainty. Furthermore, the *EC–Biotech* Panel explicitly argued that

> the European Communities has not identified possible uncertainties or constraints in the risk assessments in question, much less explained why, in view of any such uncertainties or constraints, (.) prohibition is warranted by the relevant risk assessments.[12]

So in the dispute settlement, it turned out to be vital whether or not the risk assessments identified uncertainty, either indirectly, through indicating scientific divergence, or explicitly, by stating uncertainties or constraints.

The second argument of the European Union was that the same risk assessment can 'sufficiently warrant' or 'reasonably support' more than one type of SPS measure. In other words, the European Union contended that one and the same risk assessment may justify 'divergent responses by equally responsible and representative governments'.[13] The Panel agreed with the European Union that one and the same risk assessment may serve as a basis for different types of safeguard measures, given that a risk is determined to exist. In the Panel's view, however, the risk assessments conducted by the lead CA and the EU scientific committee were favourable and contain no evidence that the biotech products subject to the case present any greater risk to human health or the environment than their conventional, i.e. non-biotech, counterparts. Yet the measures adopted by the Member States on the basis of these risk assessments provide for complete prohibition of the products.[14] Therefore, the Panel pointed out that there is no apparent 'rational relationship' between the Member States' safeguard measures, which impose a prohibition, and the risk assessments, which found no evidence that the biotech products in question are likely to cause adverse effects on human or animal health and the environment.

In conclusion, the Panel stated that the EU Member States' safeguard measures at issue were not based on a risk assessment. This conclusion highlights the key political role of the risk assessments by the lead CA and the EU scientific committees. In the cases considered, the risk assessments provided so-called 'plausibility proofs' (Van Asselt and Vos 2006, 2008; Van Asselt *et al.* 2008), i.e. conclusive statements on whether there is a risk. From a political perspective, the CA and the EU scientific committees provided zero-risk statements. The risk assessment did not involve any discussion about uncertainties and the possible or actual divergence within the scientific community about uncertain risks. In that way, the risk assessments of the lead CA and the EU scientific committees actually dictated policy. With uncertainty intolerant risk assessment suggesting full scientific consensus within the WTO regime the only science-based measure is approval.

Identifying Uncertainties in the Risk Assessment

The Panel does recognize that different governments may decide to manage risk differently and they would have the flexibility to impose different safeguard measures based on the same degree of risks identified in a risk assessment. The absolute prerequisite for a government, however, is that a risk assessment considers (uncertain) risks to human health or the environment possible. The Panel also confirms that if there are any uncertainties affecting scientific judgment, a Member can take this into account when it determines the measure to be applied in order to achieve an appropriate level of protection. In this respect, the Panel agrees that the precautionary principle is a valid risk management approach. The *EC–Biotech* Panel drew attention to the *Working Principles for Risk Analysis for Application in the Framework of the Codex Alimentarius*, which state that

> [t]he report of the risk assessment should indicate any constraints, uncertainties, assumptions and their impact on the risk assessment. Minority opinions should also be recorded. The responsibility for resolving the impact of uncertainty on the risk management decision lies with the risk manager, not the risk assessors.[15]

The Codex Alimentarius Commission explicitly states that '[r]isk managers should take into account the uncertainties identified in the risk assessment and implement appropriate measures to manage these uncertainties'.[16] In other words, the risk assessors should identify uncertainties while risk managers are responsible for choosing an appropriate way to deal with these uncertainties. This provides further support to our observation that uncertainty intolerant risk assessments are political acts as they do not allow policy makers to decide how to deal with the uncertainties. Of course, when there are no uncertainties, they cannot be identified. However, when considering innovative technologies with which society has no or very limited experience, including biotechnology, there are by definition uncertainties with regard to long-term, indirect impacts on human health and the environment. Or, as Nowotny (2007) argues even more radically, innovation generates uncertainty.

In the late 1990s, there were concerns also among scientists about the risks associated with GMOs and agro-biotechnology in particular (see Jasanoff, 2005, for a sociological analysis of three decades of debate on GMOs). The choice not to discuss them in the risk assessments pre-empted the possibility to secure acceptance of precautionary measures in the global trade arena. The fact that the EU scientific committees did not identify any uncertainties in their risk assessments forecloses the possibility for the EU Member States to take a precautionary approach. In this case, the risk assessors, by not indicating any uncertainties, restricted – if not dictated – the work of risk managers. As the Panel indicated, if there are any constraints and uncertainties that have an impact on the risk assessment, they should be identified by the risk assessors. The Panel put forward the principles of the Codex Alimentarius, stating that risk assessors are not the ones that should deal with uncertainty. This task is reserved for risk managers. When scientists do not identify any uncertainties, however, the hands of risk managers are tied.

EU Risk Assessments

The Panel emphasized that the European Communities did not identify any risks or uncertainties in the assessments qualified as risk assessments by the Panel. We, therefore, decided to have a closer look at the risk assessments in the domain of agro-biotechnology conducted by the EU scientific committees until 2002 and by the European Food Safety Authority (EFSA) and EFSA's GMO Panel after 2002.

Risk Assessment by the EC Scientific Committee

An example of the way risk assessment was conducted by the EC Scientific Committees is the opinion of the Scientific Committee on Plants (SCP) regarding Bt-11 maize adopted on 30 November 2000.[17] Bt-11 was banned in Italy in 2000 and this is one of the safeguard measures part of the *EC–Biotech* dispute. SCP assessed whether the cultivation of this genetically modified maize could have any adverse effects on human health and the environment. In the opinion of the Scientific Committee on Plants, it is stated that

> there is no evidence to indicate that the placing on the market for cultivation purposes of maize line *Bt-11* and varieties derived from this line by conventional crossing with maize lines other than genetically modified ones, is likely to cause adverse effects on human health and the environment.

As part of the scientific background on which the opinion was based, the SCP referred to a number of studies performed on mammals and birds. A host of safety and environmental aspects, including potential for gene transfer, safety of the gene product (food and feed), safety of non-target organisms and resistance and tolerance issues, were assessed. The SCP concluded that Bt-11 is not likely to cause adverse effects on human health and the environment.

It is important to note that, due to regulatory provisions and practical constraints and reasons, the Committee did not perform its own assessment of risks but instead it reviewed evidence provided by the applicant company, namely the Swiss biotech company Syngenta. The SCP mentions this in its opinion:

> The weight of evidence provided by the company and available elsewhere leads the Committee to conclude that there is no significant risk to humans or livestock following ingestion of the introduced gene products.

After explicitly concluding that Bt-11 maize does not pose any risks to human health and the environment, the SCP revealed that

> The notifier has submitted a package of relevant information […] and has carried out relevant feeding trials with target animals. Data from these studies provide no indication of risk associated with the cultivation of these GM maize lines.

This means that SCP only made a review of the applicant company's assessment instead of performing its own. Hence, the SCP opinion, which was considered a qualified risk assessment in the WTO SPS sense, is not an independent risk assessment but an expert opinion informed by the safety assessment carried out by an interested party. As far as we can judge, SCP did not juxtapose the company's assessment with studies published in the scientific literature. So 'review' might even be too strong a notion to qualify what SCP actually did. This practice inevitably compromises the independence of the Committee's assessment: it is reliant on the quality of evidence provided by the company and also the willingness of the latter to provide information unselectively. It also implies that SCP in its task as risk assessor to identify uncertainty is dependent on the willingness of the company to disclose uncertainties.

SCP concluded that there is *no* evidence that Bt-11 maize is *likely* to cause adverse effects on human health and the environment. The statement that there is no evidence of adverse effects obviously has its limitations: in particular it should be read as meaning that the applicant, i.e. the biotech company producing Bt-11, did not indicate risks. It is important to note that the assessments qualified as risk assessment in the *EC Biotech* case are not independent risk assessments but (for good reasons) heavily rooted in the assessment performed by the company. It can be questioned why the so-called 'Reasons documents' that the EU Member States produced were not used in the *EC Biotech* case as a kind of counterweight to this industry bias.

In the case of T25 maize, banned by Austria and Italy, the SCP was asked to express its opinion on the scientific information provided by Austria. In April 2000, Austria prohibited the marketing of T25 maize on the grounds that the product had not been tested under realistic conditions. After examining the information provided by the Austrian authorities, the SCP concluded that this does not constitute new scientific information and thus does not change the original 'risk assessment' (or more precisely the review of the company's safety assessment) conducted on T25 maize.[18] The main preoccupation expressed in the studies submitted by the Austrian authorities was genetic transfer. After bringing to the fore studies of pollen dispersal (Emberlin *et al.* 1999) conducted in low to moderate winds, SCP states that 'if the established separation distances developed for seed production are observed, pollen transfer to adjacent varieties should be minimized'.[19] The opinion of the SCP concludes:

> In view of the remote chance of the transfer into the environment of genetic tolerance to the herbicide glufosinate ammonium from cultivated maize, there is no evidence that this transgenic maize will pose any problem to the ecology of the alpine grassland.

In the risk assessment of T25 maize, few issues are noteworthy. Firstly, despite the complaint of the Austrian authorities that in the original risk assessment the product has not been tested under realistic conditions, in its review assessment the SCP again relied on studies that are conducted in what can be considered unrealistic conditions (i.e. assuming that there will be only low to moderate winds in the alpine region). Secondly, the SCP concludes that there is a remote chance of genetic transfer therefore there is no evidence that T25 maize will pose any risk. Thus it would seem that the absence of evidence is equated with evidence of absence, although this is violating some basic principles of risk assessment. Thirdly, the same study cited by EFSA in support of its opinion, namely the study by Emberlin *et al.*, shows that pollen from maize can be dispersed over much greater distances than had previously been accepted by government science. The authors of the study expressed concerns about the lack of acknowledgement of potential pollen spread (BBC News 1999). Thus, one of the few studies cited in EFSA's opinion does not prove the authority's point. On the contrary, it brings about more ambiguity.

It is beyond the scope of this chapter to analyse in detail all risk assessments conducted by the scientific committee that were qualified as risk assessment by the *EC–Biotech* Panel. And it is anyway beyond our expertise to qualify the risk assessments in terms of content. We only describe assessment behaviour around the risk assessments that were central to the WTO Panel's Decision. The observations pertaining to the assessment behaviour of the EU scientific committees discussed above indicate that questions can and should be asked about what was used as the critical basis for the conclusion of the *EC–Biotech* Panel that the EU Member State measures are WTO-inconsistent. The risk assessments of the EU scientific committees played a vital role in the Panel's reasoning. Those assessments, however, suffer from shortcomings which have not been addressed in the same way as the shortcomings of the assessments by the EU Member States who issued safeguard measures.

Risk Assessment by EFSA

During the time span of the *EC–Biotech* case (2003–2006), there have been other attempts to invoke the precautionary principle in the context of agro-biotechnology. A case in point is the Austrian authorities which claimed to be entitled to create a GMO-free zone. The Republic of Austria challenged the Commission's decision to reject the ban on GMOs in Upper Austria.[20] Meanwhile, the European Food Safety Authority (EFSA) was established. According to Regulation 178/2002, adopted on 28 January 2002, the main responsibility of the authority is to provide scientific opinions based on risk assessment.[21] So EFSA replaced the Scientific Committees as risk assessor in the field of agro-biotechnology. EFSA is not just a risk assessment authority but also plays a role in balancing trade and risks. Stemming from the definition in EU law, the risk assessment function of EFSA is shaped by the free movement imperative. As Alemanno (2007: 210) points out, even though the authority should be committed to a 'high level of protection of human life and health', this task must be performed 'in the context of the operation of the Internal Market'.

In March 2003, on the basis of the then Article 95(5) of the then EC Treaty, Austria proposed a draft law to the Commission to prohibit the cultivation of genetically modified seeds and planting material in the province of Upper Austria. In response, the Commission requested a scientific opinion from EFSA to investigate whether the information provided by Austria contains 'new' scientific evidence in terms of risks to human health and the environment that would justify the prohibition of GMOs, including those that have been already authorized under Directive 90/220 or Directive 2001/18. The exact phrasing was:

> as to whether the information provided by Austria in the Report entitled 'GMO-free agricultural areas – Design and analysis of scenarios and implementational measures' provides any new scientific evidence, *in terms of risk to human health and the environment*, that would justify the prohibition of cultivation of genetically modified seeds and propagating material.[22]

The word 'new' could be read as meaning 'not available at the time of earlier risk assessments' but also refers to the question of whether it concerns scientific evidence that challenges earlier risk assessments used in the regulatory process. In the context of innovation, this requirement is questionable as it supposes that there is only one legitimate interpretation of the uncertainties pertaining to risks. Reinterpretation of existing insights is therefore not accepted in the regulatory process, although that is one of the ways through which science advances. So the way in which the request for EFSA's opinion was framed is not doing justice to the situation of inherent uncertainty. Furthermore, it is interesting to note that Austria chose to consider various scenarios, which is an appropriate approach in view of uncertainty (Van Asselt *et al.* 2010). This contrasts with the request of the Commission for a plausibility proof, being a conclusive statement as to whether there is a risk.

The new law proposed by the province of Upper Austria was based on the above-mentioned report entitled 'GMO-free agricultural areas – Design and analysis of scenarios and implementational measures'.[23] The report discusses the co-existence of GMOs and conventional/organic systems and subsequent management measures. EFSA deems it necessary to point out that 'as requested, EFSA has commented on issues within its remit relating to human health and the environment and *not on other issues* such as information relating to the management of co-existence'. The specific framing of the question posed by the Commission thus either confines EFSA's risk assessment only to particular issues or, conversely, allows EFSA to engage in the type of assessment behaviour referred to by Van Asselt and Vos (2008) with reference to Jasanoff (2005), as 'not transparent, but politically significant boundary work'. This boundary work, meaning the creation and maintenance of boundaries, between what needs to be considered and what not, in this case enables EFSA to evade treatment of 'uncertainty' as such. In this case, where the focus of the scientific evidence presented by Austria is on the co-existence between GM and conventional crops, the narrow scope of the framing of the question and the way in which EFSA followed and used this framing, allowed the risk assessors to dismiss the evidence forwarded by Austria.

The result is an assessment that does not identify any uncertainties or uncertain risks related to the cultivation of GMOs. However, it is relevant that EFSA nevertheless stated that it 'was not asked by the Commission to comment on the management of co-existence of GM and non-GM crops, but the Panel recognized that it is an important agricultural issue'.[24] As a consequence of the specific formulation of the request and EFSA's interpretation of it, there is no room to consider relevant scientific insights. Following EFSA's scientific opinion, the Commission rejected Austria's safeguard measure. This uncertainty intolerant risk assessment behaviour thus seriously impacted upon, if not determined, the room for the Member State in question, Austria, to exercise precautionary risk management.

In response to the Commission decision, Upper Austria and the Republic of Austria challenged the Commission decision at the Court of First Instance, alleging a breach of the precautionary principle.[25] The Court found the alleged breach irrelevant, since the conditions under the then Article 95(5)

EC were not met. The latter article provided the possibility to derogate from harmonized measures based on Article 95 EC and was interpreted to be biased towards environmental protection. According to the interpretation of the Commission, however, national measures are only allowed when they are not 'endangering the unity of the market'. The CFI reiterated the fact that the Commission adopted the conclusions of EFSA that 'the scientific evidence presented contained no new or uniquely local scientific information on the environmental or human health impacts of existing or future GM crops or animals' and stated that 'the applicants have failed to provide convincing evidence such as to cast doubt on the merits of those assessments'.[26] Regrettably, the CFI did not examine the alleged violation of the precautionary principle.

Austria appealed the case before the European Court of Justice, stating that the Court should not have overlooked the inadequacy of the risk assessment and the violation of the precautionary principle. In Austria's view, 'the Commission did not carry out a complete scientific analysis of the risks'.[27] However, the ECJ did not review EFSA's assessment, despite the confirmation given by the CFI and ECJ in previous judgments that the legality of opinions issued by an agency scientific committee may be examined by EU courts.[28] In the *Artegodan* judgments the CFI held that even though a court cannot interfere with the risk assessment itself, it can review the proper functioning of the committee, the consistency of the opinion and its reasoning.

Precautionary Measures and the WTO

Due to the fact that the EU scientific committees and EFSA concluded in their risk assessment that biotech products do not pose any risks or uncertainties for human health and the environment, it is not possible for the EU Member states to justify their safeguard measures under Article 5.1 of the SPS Agreement. Consequently, unless precautionary measures are found to comply with the requirements of Article 5.7, they will be rendered incompliant with WTO law. Article 5.7 of the SPS agreement constitutes an exception to the requirement of Article 5.1 that every SPS measure must be based on a risk assessment. Had the Panel concluded that the Member States' safeguard measures were consistent with Article 5.7, Article 5.1 would no longer be applicable and thus the Panel would not have concluded that the European Communities acted inconsistently with its obligation under Article 5.1.

There are four requirements for a WTO Member to enact a provisional measure under Article 5.7, which were established by the Appellate Body in *Japan-Agricultural Products II*:[29]

> Pursuant to the first sentence of Article 5.7, a Member may provisionally adopt an SPS measure if this measure is:

imposed in respect of a situation where 'relevant scientific information is insufficient', and;

adopted 'on the basis of available pertinent information'.

Pursuant to the second sentence of Article 5.7, such a provisional measure may not be maintained unless the Member which adopted the measure:

'seek[s] to obtain the additional information necessary for a more objective assessment of risk', and;

'review[s] the ... measure accordingly within a reasonable period of time'.

Furthermore, it has been established that these four requirements are 'clearly cumulative in nature', meaning that if one of them is not met, the safeguard measure is not consistent with Article 5.7. In the *EC–Biotech* dispute, the Panel also discussed whether the precautionary bans by the EU Member States satisfied the Article 5.7 requirements.

In this context, the *EC–Biotech* Panel explicitly referred to the statement of the Appellate Body in *EC-Hormones* that:

Article 5.7 reflects the precautionary principle, and that the precautionary principle as such has not been written into the *SPS Agreement* as a ground for justifying an SPS measure that is otherwise inconsistent with that Agreement.[30]

The quintessence of the precautionary principle is that the lack of scientific certainty should not preclude risk management. According to this principle, it is not necessary to wait for a risk to materialize before risk management measures can be issued. So the precautionary principle allows policy makers to act in view of scientific uncertainty.

Findings of the EC–Biotech Panel Regarding Application of Article 5.7

The complaining parties argued that the Member States safeguard measures on the particular biotech products failed to meet any of the four requirements set out in Article 5.7. First, they asserted that the scientific evidence is sufficient to perform a risk assessment, as the European Communities has conducted positive risk assessments on the biotech products. Second, they alleged that the safeguard measures were not adopted on the basis of 'available pertinent information'. The SCP reviewed the measures and concluded that the information provided by the Member States did not challenge the earlier risk assessments. Third, the US and its allies contended that the Member States had not sought 'to obtain the additional information necessary for a more objective assessment of risk' and that there was no record of the Member States having sought to perform a risk assessment that would support its measure on the biotech products. Finally, the United States, Canada and

Argentina alleged that neither the Member States nor the Commission had reviewed the safeguard measures within a 'reasonable period of time'.[31]

The then European Communities was in a difficult position as it had to defend safeguard measures that deviated from the market authorizations decisions at EU level. It is, therefore, not all that surprising that its reasoning is somewhat far-fetched. The EU argued that there was insufficient relevant scientific evidence. In its view, the concept of 'insufficiency' in Article 5.7 is 'relational' and should refer to matters that are of concern to the legislator. Additionally, 'insufficient' means 'insufficient' for the production of a risk assessment 'adequate for the purposes of the legislator who must decide whether a measure should be applied, provisionally or otherwise, for one of the reasons enumerated in Annex A(1) of the SPS Agreement'.

Quite some tension exists between how the EU defined risk assessment in this dispute, compared to how it dealt with risk assessments within its own internal practice, as illustrated above. In the *EC–Biotech* dispute, the European Communities characterized a risk assessment which was adequate as one which had been 'delivered by a reputable source, [which] unequivocally informs the legislator about what the risk is with a sufficient degree of precision'. In the view of the European Union: 'the higher the level of acceptable risk, the more likely it is that the legislator may conclude, within a relatively short period of time, that the scientific evidence is sufficient and that no provisional measure is therefore necessary'. And conversely, 'the lower the level of acceptable risk, the more likely it may be that the legislator may continue to consider, for a relatively long period of time, that the scientific evidence is insufficient, and that a measure is warranted'.[32] Thus, the sufficiency of scientific evidence must be evaluated in relation to the level of acceptable risk determined by legislators. The European Communities at that time, asserted that the legislators at the Member State level were entitled to conclude that scientific evidence was insufficient for their purposes.

The *EC–Biotech* Panel addressed this relational claim of the European Communities. The Panel agreed with the EU that whether the available scientific evidence is deemed insufficient to carry out a risk assessment should be established on a case-by-case basis. The *EC–Biotech* Panel recalled the clarification given by the Appellate Body in *Japan-Apples* of the phrase 'where relevant scientific evidence is insufficient'. The Appellate Body concluded that 'relevant scientific evidence' will be 'insufficient' within the meaning of Article 5.7 'if the body of available scientific evidence does not allow, in quantitative or qualitative terms, the performance of an adequate assessment of risks as required under Article 5.1 and as defined in Annex A to the SPS Agreement'.[33] The Panel also drew attention to the fact that the Appellate Body explained in *Japan-Apples* that the notion of 'insufficiency' implies a 'relationship' between the scientific evidence and the obligation to perform a risk assessment.[34]

The Panel stated further that it did not consider that the protection goals of a legislator should be relevant in determining whether the body of available scientific evidence is insufficient. The protection goals of a legislator may be relevant in order to determine which risks a Member decides to assess with the intention of regulating them, and are also relevant for determining which measures a Member decides to impose in order to achieve its appropriate level of protection against risk. However, the Panel did not see any connection between the protection goals of the legislator and the assessment of the existence and magnitude of potential risks. According to the Panel, risk assessors do not need to know a Member's 'acceptable level of risk' in order to be able to assess objectively the existence and magnitude of a risk. So the Panel dismissed EU's relational claim.

Concerning the particular measures subject to the GMO dispute, the Panel found the SCP opinions delivered during the relevant EU approval procedures – i.e. the original assessments – and the SCP opinions delivered after the adoption of the relevant Member State safeguard measures – i.e. the review assessments – to be risk assessments within the meaning of Annex A(4) and Article 5.1 of the SPS Agreement. Thus the Panel agreed with the Complaining Parties that at the time of adoption of the Member States' safeguard measure, the body of available scientific evidence allowed the performance of a risk assessment as required under Article 5.1 and as defined in Annex A(4). The Panel concluded that as one of the four requirements of Article 5.7 is not met, the measures were inconsistent with this provision. Consequently, the Panel found that the measures at issue are also inconsistent with Article 5.1 of the SPS Agreement.

In our view, it is not so surprising that the Panel dismissed EU's relational claim, as it assumed that risk management is framing risk assessment. This stance is at odds with the current regulatory arrangements, not just at the WTO level but also in the regulatory framework at the European and Member State levels. It could be argued that the relational claim put forward brings with it the obligation to fundamentally re-design risk regulation. It is beyond the scope of this particular chapter to discuss alternative models which would be doing justice to the 'risk-management-frames-risk-assessment' view.

What we would like to draw attention to here is the fact that the above reasoning implies that doing a risk assessment is a political act, as it communicates that there is sufficient scientific evidence to do a risk assessment. So as soon as any body that is considered reputable enough has carried out a risk assessment, it seems no longer possible to satisfy the insufficiency requirement. Van Asselt and Vos (2006) describe a case where it was explicitly stated in the experts' assessment that there was too much uncertainty to carry out a risk assessment but nevertheless the same assessment was used and portrayed as risk assessment and this approach even held ground before the Court. Janssen and van Asselt in this Volume furthermore demonstrate that particular expert assessment with the 'too much uncertainty' statement even served as a reference for risk assessment in other cases. This seems to suggest that as soon as experts write something about the issue of risk, this may be treated as a risk assessment, which would also render the insufficiency argument no longer valid. We concluded already that performing uncertainty intolerant risk assessments in the context of innovation is a political act that pre-empts the possibility of precautionary risk management. However, it seems that we even have to conclude that in the WTO regime any risk assessment is a political act that impedes precautionary measures.

Further Concerns

In this context, it seems relevant to point out that the *EC–Biotech* Panel argued that 'even if a Member follows a precautionary approach, its SPS measures need to be "based on" (i.e., "sufficiently warranted" or "reasonably supported" by) a risk assessment'.[35] In other words, although the precautionary principle is reflected in Article 5.7 of the SPS Agreement, this principle does not take priority over the requirement that Members base their measures on a proper risk assessment, as prescribed in Article 5.1 (which was also the stance taken by the Appellate Body in the *EC–Hormones* case[36]). That would suggest that a risk assessment is needed to defend precautionary measures, while at the same time the risk assessment will be used as ground to argue that the insufficiency condition is not fulfilled. This presents a conundrum whereby Article 5.7, which is interpreted as reflection of the precautionary principle, can no longer be satisfied. If this interpretation is correct, that would imply that in practice it would be impossible for a government to apply precaution in a way that complies with WTO law. This would also render statements that the precautionary principle is reflected in Article 5.7 of the SPS Agreement unfounded.

The particular framing of Article 5.7 and the interpretation of 'insufficient' information are slightly worrying. 'Insufficient' is described as when 'the body of available scientific evidence does not allow, in quantitative or qualitative terms, the performance of an adequate assessment of risks'. This definition of 'insufficient' *de facto* implies that in cases where a risk assessment has been conducted already, the possibility that a measure will qualify as a precautionary one is ruled out. The difficulty, if not impossibility, to satisfy the requirements of Article 5.7 means that it is complicated for governments to get a precautionary approach accepted at the WTO level. Taking into consideration the fact that virtually all companies carry out risk/safety assessments for their products, chances are extremely slim that there could be an instance where scientific evidence will be qualified as insufficient in the WTO context.[37]

The first requirement for adopting precautionary action is insufficient scientific evidence, which proves to be the most controversial one. In the *Japan-Apples* Case, Japan argued that, in spite of the large amount of scientific evidence available, there was 'unresolved uncertainty' for which science could not provide any solution. The Panel, and consequently the Appellate Body, held that, since this is not a case where scientific information is insufficient, Japan could not rely on Article 5.7 to justify its safety measure.[38] The Appellate Body explicitly stated that 'the application of Article 5.7 is triggered not by the existence of scientific uncertainty, but rather by the insufficiency of scientific evidence'. The Appellate Body asserted that it cannot endorse an 'approach of interpreting Article 5.7 through the prism of "scientific uncertainty"'.[39]

In view of the above, the prospect of applying the precautionary principle is severely limited under WTO law and the SPS Agreement in particular. By ruling out scientific uncertainty from the scope of Article 5.7, the Appellate Body closes the door for precautionary action in cases where scientific information is available, but uncertainty reigns. Given the volume of scientific studies examining the effects of GMOs on human health and the environment, it is highly probable that the mere availability of these studies is already interpreted as disqualifying the insufficiency argument, regardless of whether a risk assessment is carried out for the particular GMO. The conclusions of the Panel and Appellate Body give serious grounds for concern. Uncertainty, which is universally recognized as the core of the precautionary principle, is ruled out of the single provision in the SPS Agreement which is considered to epitomize the precautionary principle.

Conclusions and Discussion

The *EC–Biotech* Panel adopted a persistent approach of strictly demarcating risk assessment and risk management, and prescribing each with a set of allowed actions. In the view of the Panel, risk assessors are those who should identify uncertainties and issues that might be of concern to the regulator, while risk managers are responsible for choosing an appropriate way of dealing with these uncertainties. The unwillingness of the EU Scientific Committees and EFSA to discuss constraints and uncertainties related to biotech products, however, predetermined the fate of the risk management process. By not indicating uncertainties and by equalizing *low* risk with *no* risk, the risk assessors foreclosed the room for a proper risk management strategy, where risk managers could decide on a measure to insure an appropriate level of protection. In the GMO case, the hands of the regulatory authorities of EU Member States were tied. With a risk assessment that does not identify any risks and uncertainties whatsoever, the role of risk managers becomes rather limited because safeguard measures are interpreted as illegitimate barriers to trade. With an uncertainty intolerant risk assessment, the controversial products cannot be banned or restricted in other ways.

Considering the above-mentioned issues, it can be argued that the *EC–Biotech* Panel took a step backwards from the more flexible approach adopted by the Appellate Body in *EC–Hormones*, where the adjudicators recognized the danger of drawing strict lines between risk assessment and risk management and the downsides of applying a restrictive notion of risk assessment. For the purpose of responsibly dealing with (uncertain) risks, it does not seem feasible nor appropriate, to separate the process into an exclusively technical phase and a political phase. This is also recognized in the new working principles of Codex Alimentarius, which recognizes that the 'interaction between risk managers and risk assessors is essential'. The stringent separation between the different realms of risk regulation, in which risk assessment is the obligatory point of passage before risk management, has serious downsides. The apparently technical assessments of risk are in fact opaque political acts. In the *EC–Biotech* case this had the effect of distorting the process towards risk governance, with risk assessors and risk assessments having disproportionate weight to determine the outcome of the case.

In principle, the findings of the Panel on this matter are consistent with internationally recognized standards, which state that it is the task of the risk manager to deal with uncertainty and not that of the risk assessors. This means that, in principle, whether a precautionary approach can be undertaken is conditional on the work of risk assessors, and in particular their willingness to identify uncertainties. In instances where risk assessors are uncertainty intolerant, as we saw in the GMO case, risk managers will hardly get a chance to address uncertainty and employ the precautionary principle. The *EC–Biotech* Panel seemed to be willing to recognize a precautionary approach as a valid risk management strategy. It confirmed that Article 5.7 of the SPS Agreement reflected the precautionary principle. However, the measure needs to be based on a risk assessment, while in its turn the availability of a risk assessment makes it difficult, if not impossible, to satisfy the insufficiency requirement. This presents an interesting conundrum whereby the use and justification of the precautionary principle within the WTO system remain a highly contentious issue.

It is of paramount importance that there is a proper interaction between the different phases of risk analysis. The *EC–Biotech* case demonstrates the threat of rupturing the link between science and policy and the repercussions this has for the effective management of risks. Once risk assessment is taken in clinical isolation from the rest of the process and its outcomes are considered as a given, it is very likely that risk assessors, particularly uncertainty intolerant ones, will foreclose the space for risk managers to adopt appropriate strategies for dealing with risks. This seems neither appropriate nor desirable.

Our analysis suggests that whether uncertainty is tolerated and how uncertainty is dealt with in the risk assessments is politically highly significant. From this perspective, the views of the Health Council of the Netherlands (Health Council of the Netherlands 2008) and the Dutch Scientific Council for Government Policy (WRR 2009) on precaution as the obligation to concentrate on uncertainty in the whole process of dealing with risks should be endorsed wholeheartedly. The history of trade conflicts surrounding GMOs suggests that it may be too late to safeguard the possibility of precautionary risk management in that context. At the same time, this history emphasizes how important uncertainty tolerance is in the early days of a new technology. So it might still be possible to protect the room for precaution for other new technologies, with nanotechnology the first in line. There should be room for a political decision whether or not precaution is an appropriate strategy. This risk management option should not be foreclosed and pre-empted by uncertainty intolerant risk assessments. The current organization of international trade simply requires uncertainty tolerant risks assessments in order to anyway open up the possibility of precaution-based risk management. So the way in which uncertainties are dealt with in risk assessments requires far more serious attention.

Notes

1 Until 30 November 2009 the European Union was known officially in the WTO as the European Communities for legal reasons.
2 Panel Report, *European Communities-Measures Affecting the Approval and Marketing of Biotech Products*, WT/DS291/R, WT/DS292/R, WT/DS293/R, adopted on 21 November 2006.
3 Agreement on the Application of Sanitary and Phytosanitary Measures, Art. 2.2.
4 *Ibidem*, Art. 5.1.
5 *Ibidem*, Annex A(4).
6 *Ibidem*, Art. 5.7.
7 Appellate Body Report, *European Communities – Measures Concerning Meat and Meat Products*, WT/DS26/AB/R,WT/DS48/AB/R, adopted on 13 February 1998, para. 124.
8 *Supra* note 1, para. 7.2534.
9 Until May 2003 there were five Scientific Committees providing the European Commission with scientific advice regarding food safety. The Committees were created by the Commission on 10 June 1997. Two of these Committees, the Scientific Committee on Plants and the Scientific Committee on Food, performed risk assessments concerning the safety of the biotech products that were at issue in the *EC–Biotech* case. As of May 2003 all five Scientific Committees were transferred to the European Food Safety Authority (EFSA). For more details about the work of the different Committees, see the website of DG Health and Consumer Protection, available online: <http://ec.europa.eu/food/committees/scientific/index_en.htm>, accessed 20 September 2009.
10 *Supra* note 1, para. 7.3012.
11 *Ibidem*, para. 7.3059.
12 *Supra* note 1, para. 7.3066.
13 *Ibidem*, para. 7.3063.
14 *Ibidem*, para. 7.3064.
15 Codex Alimentarius Commission, *Working Principles for Risk Analysis for Application in the Framework of the Codex Alimentarius*, adopted June/July 2003, Section III, Codex Procedural Manual, 14th edition, 2004, at para. 25.
16 Codex Alimentarius Commission, *Principles for the Risk Analysis of Foods Derived from Modern Biotechnology* (adopted in June/July 2003), CAC/GL 44–2003, at para. 18.
17 Opinion of the Scientific Committee on Plants on the submission for placing on the market of genetically modified insect resistant and glufosinate ammonium tolerant (Bt-11) maize for cultivation. Notified by Novartis Seeds SA Company (notification C/F/96/05–10) (opinion adopted by the Scientific Committee on Plants on 30 November 2000), available online: <http://ec.europa.eu/food/fs/sc/scp/outcome_gmo_en.html> (accessed 23 September 2009).

18 Opinion on the invocation by Austria of Article 16 of Council Directive 90/220/ EEC regarding a genetically modified maize line T25 notified by AGREVO FRANCE (now AVENTIS CROPSCIENCE, REF. C/F/95/12–07), Opinion adopted by the Scientific Committee on Plants on 30 November 2000, available online: <http://ec.europa.eu/food/fs/sc/scp/out85_gmo_en.html> (accessed 29 November 2009).
19 *Supra* note 17, p. 4.
20 CFI, Joined Cases T-366/03 and T-235/04, *Land Oberösterreich and Republic of Austria v Commission*. See OJ 2005 C 296/.
21 Regulation 178/2002 of the European Parliament and of the Council of 28 January 2002 laying down the general principles and requirements of food law, establishing the European Food Safety Authority and laying down procedures in matters of food safety, OJ 2002 L 31/1.
22 Opinion of the Scientific Panel on Genetically Modified Organisms on a question from the Commission related to the Austrian notification of national legislation governing GMOs under Article 95(5) of the Treaty, *The EFSA Journal* (2003) 1, 1–5, p. 3.
23 Report entitled 'GMO-free agricultural areas – Design and analysis of scenarios and implementational measures' by Werner Müller (document translated by Commission services from the original title 'GVO-freie Bewirtschaftungsgebiete: Konzeption und Analyse von Szenarien und Umsetzungsschritten').
24 *Supra* note 22, p. 4.
25 *Supra* note 20.
26 *Supra* note 18, I-7194.
27 ECJ, Joined Cases C-439/05 P and C-454/05 P, *Land Oberösterreich and Republic of Austria v. Commission of the European Communities*, ECR[2007] I-07141.
28 CFI, Joined Cases T-144/00, T-76/00, T-83/00, T-84/00, T-85/00, T-132/00 and T-141/00 *Artegodan and others v. Commission*, ECR[2002] II-4945.
29 *Japan – Measures Affecting Agricultural Products*, Appellate Body Report, adopted on 19 March 1999, para. 89. See also Appellate Body Report *Japan – Measures Affecting the Importation of Apples*, WT/DS245/AB/R, adopted 10 December 2003, at para. 176.
30 See *supra* note 8, para. 124.
31 *Supra* note 1, para. 7.3222.
32 The European Communities considers that this analysis is confirmed by the words 'reasonable period of time' in Art. 5.7.
33 *Supra* note 30, para. 179.
34 *Supra* note 1, para. 7.3234.
35 *Ibidem*, para. 7.3065.
36 *Supra* note 8, paras. 123–125.
37 This preoccupation is expressed by a number of authors; see Bernasconi-Osterwalder and Olivia (2006); Negi (2007); Perez (2007).
38 *Japan – Measures Affecting the Importation of Apples*, WT/DS245/R, Panel Report, circulated 15 July 2003, para. 8.221.
39 *Supra* note 30, para. 184.

Bibliography

Alemanno, A. (2007) *Trade in Food: Regulatory and Judicial Approaches in the EC and the WTO*, London: Cameron May.
Arcuri, A., Gruszczynski, L. and Pearson, H. (2010a) 'WTO: Global Governance of Risks: WTO, Codex Alimentarius and Private Standards – Report on the SRA-Europe 19th Annual Conference', *European Journal of Risk Regulation*, 3: 285–87.

Arcuri, A., Gruszczynski, L. and Herwig, A. (2010b) 'Independence of Experts and Standards for Evaluation of Scientific Evidence under the SPS Agreement – New Directions in the SPS Case Law', *European Journal of Risk Regulation*, 2: 183–88.

Van Asselt, M.B.A. and Vos, E. (2006) 'The Precautionary Principle and the Uncertainty Paradox', *Journal for Risk Research*, 9(4): 313–36.

——(2008) 'Wrestling with Uncertain Risks: EU Regulation of GMOs and the Uncertainty Paradox', *Journal of Risk Research*, 11(1–2): 281–300.

Van Asselt, M.B.A., Van 't Klooster, S.A., Van Notten, P.W.F and Smits, L.A. (2010) *Foresight in Action: Developing Policy-oriented Scenarios*, London: Earthscan.

BBC News (1999) *GM Pollen Warning*. Available online: <http://news.bbc.co.uk/2/hi/science/nature/288865.stm> (accessed 26 September 2011).

Bernasconi-Osterwalder, N. and Olivia, M.J. (2006) 'EC–Biotech : Overview and Analysis of the Panels' Interim Report', available online: <http://www.ciel.org/Publications/EC_Biotech_Mar06.pdf> (accessed on 1 June 2009).

Emberlin, J., Adams-Groom, B. and Tidmarsh, J. (1999) *A Report on the Dispersal of Maize Pollen*. Online, available online: <http://www.mindfully.org/GE/Dispersal-Maize-Pollen-UK.htm> (accessed 28 September 2011).

Flett, J. (2010) 'If in Doubt, Leave it Out? EU Precaution in WTO Regulatory Space', *European Journal of Risk Regulation*, 1: 20–30.

Health Council of the Netherlands (2008) *Prudent Precaution*, The Hague: Health Council of the Netherlands, 2008; publication No. 2008/18E.

Janssen, A.-M. and Van Asselt, M.B.A. (2013) 'The Precautionary Principle in Court: An Analysis of Post-Pfizer Case Law', in M.B.A. van Asselt, E. Versluis and E. Vos (eds), *Balancing Between Trade and Risk: Integrating Legal and Social Science Perspectives*, Abingdon: Taylor & Francis.

Jasanoff, S. (2007) *Designs on Nature: Science and Democracy in Europe and the United States*, Princeton, NJ: Princeton University Press.

Millstone, E. (2013) 'Science and Decision-making: Can we both Distinguish and Reconcile Science and Politics?', in M.B.A. van Asselt, M. Everson and Ellen Vos (eds), *Trade, Health and the Environment: The European Union put to the Test*, Abingdon: Taylor & Francis Books.

Negi, A. (2007) 'World Trade Organization and the EC Biotech Case: Procedural and Substantive Issues', *International Studies*, 44(1): 1–22.

Nowotny, H. (2007) 'Introduction: The Quest for Innovation and Cultures of Technology', in H. Nowotny (ed.), *Cultures of Technology and the Quest for Innovation*, Oxford: Berghahn, 1–23.

Perez, O. (2007) 'Anomalies at the Precautionary Kingdom: Reflections on the GMO Panel's Decision', *World Trade Review*, 6: 265–80.

Sampson, G. (2001) 'Risk and the WTO', in T. Robertson and A. Kellow (eds), *Globalization and the Environment. Risk Assessment and the WTO*, Cheltenham: Edward Elgar Publishing, 15–27.

WRR, Scientific Council for Government Policy (2009) *Uncertain Safety. Allocating Responsibilities for Safety*, Amsterdam: Amsterdam University Press.

Part II

Risk Regulation in the European Union

5 Between Politics and Science

Accommodating National Diversity in GMO Regulation

Vessela Hristova

Introduction[1]

The motto of the European Union, *Unity in Diversity*, reveals a fundamental tension in the structure of the EU and simultaneously hints at how this tension might be resolved. The desire of the EU States to achieve common goals such as prosperity, peace and security, drives forward integration, a process that in many cases demands the replacement of particular national policies and practices with commonly agreed EU ones. At the same time, as the EU reaches deeper into historically established domestic political balances, national institutional arrangements, societal norms and cultural values, progress toward integration is stalled unless some room for toleration of national diversity is provided. This reveals a paradox at the heart of EU integration: it requires that, at some times and in some ways, national diversity be accommodated.

A focus on the mechanisms through which the EU accommodates national diversity allows us to explore analytically the fault line between the pressures for EU unity and counteracting pressure for maintaining local specificities. Particularly relevant is to identify what types of mechanisms are used within the regulatory framework to reconcile these differences, how often they are used, and to what effect. It is equally important to analyse why some requests for such accommodation are granted while others are denied.

This chapter applies the above questions to the case of agricultural biotechnology in the EU, where the tensions between unity and diversity are thrown into sharp relief. On the one hand, the single EU market exerts strong pressures toward uniformity by requiring, specifically, the setting of a common risk level for biotechnology. On the other hand, Member States exhibit important differences in their interpretations of available scientific evidence, in socio-economic circumstances, agricultural policy objectives, and levels of societal risk acceptance.

The study draws on thirteen unstructured elite interviews conducted since June 2009 with officials from the European Commission, the European Food Safety Authority (EFSA) and national delegations to the EU, as well as experts in national administrations and stakeholders such as representatives of agribusiness associations. It further relies on official documents such as Commission proposals, communications, working documents, Council minutes, and stakeholders' position statements as well as on the secondary literature.

The chapter proceeds in the following way. The next section elaborates some features characteristic of the regulation of agricultural biotechnology in the context of the EU and provides an overview of the regulatory framework for GMO authorisations. Next, two modes of diversity accommodation are introduced and their functioning with regard to national claims of scientific and non-scientific nature is evaluated. The last section provides an account of recent national initiatives to reform the framework and the European Commission's response.

Regulation of Agricultural Biotechnology in the EU: An Overview

A Complex Policy Area

The regulation of agricultural biotechnology is one of the most contentious issues in EU policy making. Several characteristics of the GMO regulation make it particularly complex: it is a cross-sectoral and multi-level issue (Pollack and Shaffer 2010); it concerns risk regulation under uncertainty which raises questions about the relative role of science and politics in the policy making process (Everson and Vos 2009); finally it is an issue area of high salience. Each of these features is briefly discussed below.

First, GMO policy concerns and carries implications for multiple sectors such as external trade, Internal Market, agriculture, enterprise, research and development, environment, consumer protection and public health. The number and variety of groups that seek to influence the policy making process such that their – often conflicting – interests are reflected in the final regulatory outcome, make that process immensely complex and the reaching of a mutually acceptable agreement difficult.

Second, GM policy making is influenced by processes unfolding at several levels: sub-national, national, EU and international. Most aspects of agricultural biotechnology are formally regulated at the EU level but the policy outcomes are determined, at least in part, by processes taking place at the national level. In the absence of a meaningful EU public sphere, for example, Member States remain the arena where public preferences on this issue are aggregated. In addition, the obligations arising from the EU's membership in the WTO place constraints on EU policy concerning trade in GM goods. Products consisting of or containing GMOs that have not been authorised at EU level cannot be imported on to the Internal Market, which raises concerns about EU authorisation requirements functioning as non-tariff barriers to international trade.

Third, GMO regulation not only spans multiple sectors and levels but also involves a complex interplay between science and politics. The highly technical nature of the issue requires reliance on scientific expertise. At the same time, GM policy is a case of risk regulation under scientific uncertainty. Actual or potential direct or indirect effects of GM technology on the environment as well as human and animal health have not yet been determined conclusively. The presence of uncertainty makes policy making based exclusively on science problematic, and complicates the interplay between science and politics in the policymaking process. Joerges (2001: 2) aptly captures the dilemma: 'the "law" cannot resolve the cognitive dimension of risk; "science" cannot provide answers to the normative dimensions'.

Finally, GMO regulation is a salient issue. The BSE food crisis in the mid- to late 1990s discredited EU food safety regulation and drew the attention of consumers to the deficiencies in EU governance. In addition to the BSE and various other food scandals, the activities and campaigns of environmental and consumer non-governmental organisations have ensured that GMO regulation remains an issue that continuously attracts popular attention and is high on the consumer's radar. Thus the general public, typically disinterested with the highly technical subject matters of EU policy making, exhibits heightened interest in GMO policy making. Such a level of public engagement presents the EU with a relatively new context for policy making that further reinforces the complexity of the science–politics nexus.

Development of the Regulatory Regime for Genetically Modified Organisms

The EU first entered the area of biotech regulation in the early to mid-1980s with the adoption of non-binding measures, the main objective of which was to harmonise the rules for laboratory research in order to ensure a level playing field for biotech companies across the EU.[2] In the early 1990s, the Council passed the first binding directives regulating the contained use of genetically modified micro-organisms (laboratory research)[3] and their deliberate release into the environment (field trials and their placing on the market for use in food and feed, industrial purposes and cultivation).[4]

The first authorisations for GMOs based on this initial regulatory framework were adopted in 1996. They coincided with a number of developments that in 1998 led the Member States to suspend all authorisations of GMOs, thus imposing a *de facto* moratorium, until the framework could be reformed in such a way as to apply stricter rules, in particular on the labelling and traceability of GMOs. Some of these developments included a highly visibly negative campaign against GM products that accompanied the first arrival of GM products on EU territory and the BSE food crisis that swept across the EU in 1996. The latter seriously undermined public confidence in the EU's scientific risk assessments and, more generally, in the capacity of the EU to regulate food safety successfully. It also turned food regulation from a relatively technical into a salient issue (Chalmers 2003).

In the wake of the BSE debacle, the Commission undertook a series of regulatory reforms in the area of food safety that applied to GM authorisations as well. Some of the main changes included a new approach to risk analysis, the establishment of the European Food Safety Authority (EFSA) and heightened attention to the precautionary principle. Following the approach proposed in the Commission's *White Paper on Food Safety* from 1999,[5] risk analysis was broken down into three separate tasks: risk assessment, risk management and risk communication. The change indicated the understanding that these tasks involve different rationales and hence require separate sets of actors to deal with each of them. Risk assessment covers the scientific and technical aspects of the regulatory process while risk management concerns the political dimension of regulation and control.

Reflecting the changes in the EU's approach to risk analysis, Regulation (EC) 178/2002 (General Food Law)[6] established the European Food Safety Authority, an independent scientific body at EU level charged with performing the risk assessment part of the overall risk analysis, including the risk assessments for GMO authorisations.[7] EFSA itself is not invested with decision-making authority. Instead it provides scientific opinions that serve as the basis for decisions formally adopted by the Commission through the comitology procedure.[8]

Finally, the precautionary principle, which previously was formulated only through judgments of the European Court of Justice,[9] is now formally codified in the General Food Law.[10] It allows the use of provisional measures in cases of scientific uncertainty where a possibility of harm to health exists, even if it is not yet conclusively proven scientifically. According to the Commission's Communication on the *Precautionary Principle*, 'Judging what is an "acceptable" level of risk for society is an eminently *political* responsibility. Decision-makers faced with an unacceptable risk, scientific uncertainty and public concerns have a duty to find answers'.[11] This quote reveals the awareness of the Commission of the limits of science in the policy making process. It underlines an understanding that science only is insufficient to provide all and complete answers to regulatory questions.

In addition to the overall restructuring of the food safety regime in the EU, which carried implications for biotech policy making as well, the Commission also undertook reforms that targeted the biotech regulatory framework. These efforts aimed to address the concrete concerns that by 1998 had led Member States to impose the moratorium on GMO authorisations and require stricter rules. The reform was finalised in 2003 and the authorisation process resumed with the first GMO approvals adopted in 2004. In specific Directive 2001/18 replaced Directive 90/220 as the framework legislation that regulates the non-contained release of GMOs, i.e. beyond the laboratory. The 2001 Directive sets out harmonised procedures for the authorisation of GMOs for two purposes: experimental release (field trials) and placing GMOs on the market for cultivation, use in food and feed, or industrial processing. Subsequently, Regulation 1829/2003 further updated the rules for GMO authorisations for use in food and feed in particular. Regulation 1830/2003 set out requirements for labelling and traceability and harmonised thresholds for the adventitious and technically unavoidable presence of GMOs in products.[12] The reformed regulatory package introduced improved environmental risk assessment, paying attention to the long-term effects; post-marketing monitoring; temporally limited authorisations subject to renewal; abolition of the simplified procedure for GMOs for use in food, and feed; and the abandonment of the substantial equivalence principle. Finally, the Commission has put forward a non-binding Recommendation[13] for co-existence measures in favour of farm-level management measures.

The GMO Authorisation Procedure

The rules that govern the process of GMO authorisations have two main objectives: ensuring the smooth functioning of the Internal Market with regard to GMO products and guaranteeing a high level of protection of the environment and human and animal health. In contrast to the US product-oriented approach, the EU has adopted a case-by-case, process-oriented approach.[14] This means that each GMO variety has to undergo individual risk assessment and must be authorised before it can be placed on the market.

The authorisation procedure for the marketing of GMOs did not fundamentally change in the reformed framework.[15] The procedure, under the 2001 Directive, for authorisations of GMOs intended to be placed on the market functions in the following way. A company submits an application (notification) to the competent national authority (CNA) of the country where the product is first to be marketed. The application follows a standardised format set out in the relevant EU legislation. The national authority checks for completeness of the dossier and performs the risk assessment. It communicates the results to the European Commission and the Member States, stating whether the GM variety should be authorised or not. Member States and the Commission have a period of time in which they can make comments, request further information or raise objections. If the objections cannot be resolved bilaterally the issue is transferred to the European level and handled through a Community [EU] procedure. In accordance with it, the European Food Safety Authority issues its own scientific opinion on the matter. Then the Commission, taking into account EFSA's risk assessment report, proposes a decision on the authorisation of the GMO which is put to a vote in front of a regulatory committee composed of national representatives: the Standing Committee on the Food Chain and Animal Health (SCFCAH). If the committee cannot reach a qualified majority to accept the proposed decision, the draft is forwarded for a vote in the Council. If the Council fails to act within the specified timeframe or cannot reach a decision, the Commission adopts its decision. In the case of a decision in favour of authorising a GMO, the national competent authority of the country where the application was originally submitted issues the final consent to the notifier.[16]

The remainder of this chapter examines how this formal procedure works in practice and the extent to which it is capable of accommodating national divergent opinions.

Accommodation of National Diversity in GMO Authorisations

Two Modes of Diversity Accommodation

As noted earlier, agricultural biotechnology cuts across multiple sectors and touches upon a wide range of issues. As a result, Member States have developed positions informed by a broad set of considerations, including divergent level of risk acceptance about health and environment, national objectives for the development of certain types of agricultural production or for stimulating innovation and research, socio-economic considerations, and public attitudes toward genetic engineering. How then does the EU regulatory framework mediate and accommodate these diverse local preferences that are often conflicting with the universal logic of science and an integrated EU market?

In thinking broadly about the ways in which national diversity can be accommodated, two general modes are distinguishable: accommodation through deliberation and accommodation through differentiation. Under the first mode, accommodation happens during the decision-making process at the EU level and is reflected in the substance of the policy outcome. According to this mode, the full range of considerations that inform the national positions of Member States are voiced and duly taken into account during the decision-making process.[17] Such a mode of accommodation presupposes a regulatory framework that is open to a wide range of considerations. As a result, under this mode of accommodation, one should be able to observe a variegated set of policy outcomes that do not systematically privilege one type of argument over others. Conversely, outcomes consistently of one type might indicate that the regulatory framework is biased toward one type of consideration to the systematic exclusion of another, thus not providing for adequate accommodation of diversity through deliberation.

A second mode of accommodation allows individual Member States to deviate from the commonly agreed solution rather than attempt to work their preferences into the final outcome. Whereas the first mode affects the outcomes of the decision-making process at the EU level, such as the common level of protection, the second mode accommodates diversity through differentiation. It allows some Member States to deviate from the policy outcome and implement a different one that is closer to their respective national positions. Thus, rather than having one policy outcome that reflects a wide variety of types of concerns and holds equally for all Member States, the second mode of accommodation allows for multiple outcomes through opt-outs or other flexibility measures.

The following sub-section applies these two conceptualisations of accommodation to the case of GMO authorisations in the EU. It analyses the extent to which two types of considerations – the scientific and non-scientific – can be and are accommodated under the two modes outlined above. While Member States certainly hold scientific considerations as important, their national positions are formed as a result of pressures that go beyond science. Therefore, analysing the admissibility of scientific *versus* non-scientific considerations is one way to identify opportunities for accommodation of national diversity.

Accommodation Through Deliberation

For national diversity to be accommodated through deliberation, the regulatory framework needs to be open to a diverse range of considerations, including those both scientific and non-scientific in nature. The following section analyses the extent to which this is the case in the process of GMO authorisation.

The division of risk analysis into the separate tasks of risk assessment and risk management has the purpose to assign clear and specific responsibilities to different sets of actors and thus provide more transparency compared to the time before the BSE crisis. Scientists are tasked with producing scientifically grounded opinions while the final decision-making authority resides with policy makers whose role involves making normative judgments about acceptable levels of risk based, in part, on the technical information provided by the risk assessors. Recital 32 of Regulation 1829/2003 acknowledges the distinct role of risk managers. It explicitly recognises that science cannot be the sole basis for decision making and that risk management involves considering additional kinds of rationale that go beyond the scientific evaluation of risk: 'in some cases, scientific risk assessment alone cannot provide for all the information on which a risk management decision should be based, and that other legitimate factors [...] may be taken into account'. Article 19 of the same Regulation specifies the elements upon which the Commission may base its proposal for an authorisation decision and they include 'the opinion of the Authority [EFSA], any relevant provision of community law and *other legitimate factors* relevant to the matter under consideration' [Italics added]. From this it appears that the regulatory framework formally provides some basis for the Commission to go further than the scientific opinion of EFSA. Thus, formally, both scientific and non-scientific considerations can find expression in the decision-making process.

Despite the leeway to do so, in practice the Commission has never included non-scientific factors in its decisions and has always followed the opinions of EFSA. EFSA, for its part, has always found that GMOs are safe and recommended their authorisation (Kritikos 2009: 424). There has been only one case (November 2007) in which DG Environment, in charge of the dossier at the time, drafted a proposal that deviated from EFSA's opinion against the authorisation of the Amflora potato. This draft proposal for approval was struck down at an internal Commission meeting and was never formally adopted as a proposal to be put to a vote in the Regulatory Committee. It is possible that this was a calculated move on the part of the Environment Commissioner who anticipated that the draft proposal would not pass the College of Commissioners, but wanted to send a positive signal to the public and the environmental organisations (Interview with Commission Official). Apart from this aborted attempt to go beyond EFSA's opinion, the Commission has always based its proposals on the recommendations of its scientific agency and never gone over or beyond them. Neither has the Commission ever included in its decisions references to non-scientific concerns. In short, there is a mismatch between the legislative framework with its – albeit peripheral – openness to non-scientific concerns and the 'measurable and quantifiable forms of argumentation' that the Commission has given preference to in its implementation of the framework (Kritikos 2009: 425). It has not made use of the full range of options that the regulatory framework allows. A slightly different interpretation assigns the responsibility for the fact that the decision-making process is mostly based on science to the structural division of the risk analysis into scientific risk assessment and political risk management (Chalmers 2005). It puts the Commission under the obligation to provide reasons for deviating from EFSA's scientific opinions and thus elevates scientific knowledge to a privileged position.

In some cases, national representatives in the Standing Committee on the Food Chain and Animal Health have succeeded in modifying the Commission proposal to include more stringent post-marketing monitoring measures than the general surveillance proposed by EFSA (Interview with national representative). However, these modifications in the risk management measures too have been based on scientific information, conducted by national competent authorities. In the risk management phase the debate between Member States' representatives and the Commission continues to be science-based. The fact that risk management too is based exclusively on scientific rationales makes doubtful the meaningfulness of the distinction between a scientific assessment phase and a political risk management phase. Similarly, whenever the Commission has delayed the adoption of an authorisation decision, it has done so in order to request or consult further scientific studies. Rather than working out procedures allowing the expression of non-scientific concerns such as divergent societal levels of risk acceptance, the Commission has preferred to eschew the issues by appealing to the universalism of science.

In short, the risk management phase of the authorisation process in practice has little added value above and beyond the scientific risk assessment performed by EFSA. By under-utilising its risk management role, the Commission makes the authorisation process exclusively based on science and does not take advantage of possibilities for accommodation of non-scientific concerns through the decision-making process. Member states bring in scientific argumentations (especially in their submissions to EFSA during the risk assessment phase) but in addition their national positions are formed as a result of broader set of considerations, such as societal levels of risk tolerance, public opinion, the structure of the agricultural sector or objectives for its development. When the latter are not even admissible in the decision-making process, the result is reduced chances for some Member States to have their positions accommodated.

Accommodation Through Differentiation

The previous section has argued that while the regulatory framework for GMO authorisation formally allows for the consideration of both scientific and non-scientific factors, the way the application of the framework works in practice leaves little space for non-scientific considerations from Member States. What remains to be seen is how and according to what criteria the second mode of accommodation functions, whereby national considerations are taken into account in the post-authorisation stage and allow deviation from EU authorisation decisions. There are two accommodating mechanisms in particular that fall under this mode: a directive-specific safeguard clause (as indicated by Article 114(10) TFEU) and a general treaty-based derogation procedure (opt-out) laid out in Article 114(4) and (5) TFEU (ex Article 95(4) and (5) EC). Each of these mechanisms is introduced and discussed below.

Safeguard Clause

Article 23 of Directive 2001/18 and Article 34 of Regulation 1829/2003 – safeguard clauses – present one form of accommodation through differentiation whereby Member States can deviate on a case-by-case basis from the result of a specific authorisation procedure. The safeguard clause under the 2003 Regulation has not been used so far and the analysis below focuses on the instrument available under the 2001 Directive.[18] Article 23 of Directive 2001/18 allows an individual Member State to ban a GMO that has been officially authorised at EU level from being placed on its national market in cases when it deems that the GMO presents risks to human health or the environment. The Member State needs to provide reasons for its decision in the form of new and additional information that has become available after the GMO approval such that it would affect the conclusions of the scientific risk assessment on which the authorisation was based. The ban is provisional but effective immediately upon notification to the Commission. After receiving the national notification of the placement of the ban, the Commission then examines the argumentation of the deviating Member State. It can consult EFSA on the merits of the scientific evidence provided by the Member State in justification of its ban. After EFSA issues its opinion, which typically concludes that the safeguard measure is scientifically unfounded, the Commission proposes a decision to lift the ban according to the familiar comitology procedure followed for authorisations.

There are 13 national bans currently in place, based on the safeguard clause in the 2001 Directive. Seven countries (Austria, Bulgaria, France, Germany, Greece, Hungary and Luxemburg) have at different times introduced a ban on the use for cultivation purposes of GM maize variety MON810. Austria, Hungary and Luxemburg have already placed a ban on the cultivation of the most recently authorised Amflora potato (written communication with Commission Official, 2011). In addition, Austria has three more bans: one on the cultivation of GM maize T25, and two for non-cultivation purposes (MON836 and GM oilseed rape products). To the extent that MON810 maize and the Amflora potato are the only two GMOs on the market effectively authorised for cultivation (T25 is not actively marketed by its company, Bayer), the national bans by Austria, Hungary and Luxembourg amount to a blanket ban on GMO cultivation on their respective territories.

The Commission has repeatedly attempted to overturn the bans, relying on EFSA's opinions regarding the merits of the scientific evidence provided by the national governments.[19] However, in this respect it has encountered a Council that is more capable of reaching qualified majority (QM) than when it comes to authorisation decisions. First, Member States are naturally more hesitant to vote against a fellow EU country forcing it to lift its national ban as politically this is a very sensitive matter. Furthermore, besides the practical uneasiness of voting against a fellow government, some Member States, even if they are pro-GMO, vote against the Commission proposal because they believe that Member States should have the capacity to impose restrictions on the cultivation or marketing of GMO on their territory (Pollack and Shaffer 2010: 352, Skogstad 2011). Such countries vote for the principle of preserving national competences rather than on substantive grounds. Thus, in contrast to GMO approvals, when it comes to lifting safeguard bans, final decisions have deviated from the scientific opinions of EFSA to *de facto* reflect a broader scope of factors, such as political considerations about the suitable division of competences between the EU and Member States.

The Commission has, nonetheless, succeeded on two occasions in making a Member State lift its safeguard measures. In June 2005, it proposed a decision challenging collectively all national bans. This proposal was overwhelmingly rejected in the Council.[20] Next, the Commission targeted just one Member State (Austria) with regard to their bans on two GM crops for the purposes of both cultivation and use in food and feed. These bans were also upheld with only the Czech Republic, Netherlands, Sweden and the UK supporting the Commission (Pollack and Shaffer 2009: 259). Finally, in October 2007, the Commission proposed a decision with even narrower scope for lifting the ban just on non-cultivation use (i.e. in food or feed) of the two GM varieties banned by Austria. This time, the Council failed to reach the QM needed to reject the Commission decision. Thus, the final word was left to the Commission and in May 2008 it adopted a decision asking Austria to lift its bans, a decision with which Austria complied. Most recently in

March 2009, the Commission again tried to lift the ban on cultivation for Austria and Hungary but the Council rejected that proposal.

The use of the safeguard clause has produced more variegated outcomes, with some bans being upheld by the Council and others ordered by the Commission to be lifted. Moreover, the outcomes from the use of the safeguard clause have deviated from EFSA opinions in contrast to authorisation decisions. To the extent that national bans are more often retained contrary to EFSA's scientific opinions and Commission proposals, one may conclude that in practice safeguard measures reflect non-scientific considerations and provide greater flexibility to accommodate Member States compared to the authorisation procedure. This is due to the fact that reaching a QM in the Council regarding safeguard measures of a specific Member State is easier than on general authorisation decisions. However, maintaining safeguard measures is by no means an automatic and uncontested process guaranteeing Member States an easy exit from Community obligations. The ban is subject to scrutiny by the Commission and EFSA and can be struck down. While there is a certain sense of solidarity among Member States in the Council, their support cannot be taken for granted. Member States wishing to retain their bans actively lobby the other Member States for support in order to secure a qualified majority vote against the Commission proposals (Pollack and Shaffer 2009: 259).

The safeguard clause is really meant as a safety valve for emergency or crisis situations where immediate reaction is needed to prevent threats to human health or the environment rather than as a way to accommodate national differences. In practice, however, the application of the safeguard clause comes closer to functioning as a mechanism for diversity accommodation due to the fact that the procedure involves voting by national executives who are more sensitive to political and socio-economic considerations when it concerns decisions of individual Member States rather than GMO authorisations in general.

Article 114(4) and (5) TFEU

Article 114(4) and (5) TFEU (ex Article 95(4) and (5) EC) formally offers another route for individual Member States to deviate from the commonly adopted decisions on authorisations of GMOs, and thus is a part of the second mode of accommodation. Article 114 TFEU (also known as the Internal Market clause) provides the legal basis for harmonisation of national legislation at the EU level in cases where differences in national legislation obstruct the functioning of the Internal Market by creating non-quantitative barriers to the free movement of goods. Paragraph (4) of Article 114 TFEU allows a Member State to maintain its already existing national measures if they provide a higher level of protection. Paragraph (5) of the same Article allows a Member State to introduce new national measures that deviate from the EU legislation if it wishes to apply more stringent provisions than those set out in the harmonised legislation. As both the Deliberate Release Directive and Regulation 1829/2003 are legally based on the Internal Market clause, Article 114(4) and (5) TFEU is available as an opt-out mechanism. Paragraph (4) has not been used so far as a basis for a national ban and therefore the rest of the analysis focuses on cases relying on Paragraph (5).

While the safeguard clause in the 2001 Directive discussed in the preceding section allows bans on individual GMO authorisations, recourse to Article 114(5) TFEU could produce more general results (Fleurke 2008). First, the exemption is not necessarily valid for only a limited period of time. Second, it could imply a ban on all GMOs rather than on a case-by-case basis because it could secure opt-out from the legislation as a whole.

The conditions for relying on this Article are somewhat more restrictive than those in the safeguard clause. Similarly to the safeguard clause, a national government needs to provide new and additional scientific information that has become available after the adoption of the legislation. In addition, it also needs to demonstrate that what follows from this new scientific knowledge uniquely affects the population of the country seeking the opt-out. In short, recourse to this clause is again admissible only on scientific grounds. An important procedural difference, however, is that the opt-out does not enter into force immediately upon notification as is the case with the safeguard clause. The opt-out request first needs to be examined and approved by the Commission. Again in contrast with the safeguard clause where the review of the national safeguard ban is subject to the comitology procedure, in the case of Article 114(5) TFEU, the Commission alone has authority over the approval of the opt-out request.

Several countries – Austria, Poland and most recently, Portugal – have tried to employ this mechanism in an attempt to adopt national laws establishing GM-free zones at a provincial level.[21] Because establishing a GM-free zone implies a blanket ban on all GMOs, it constitutes a more stringent level of protection and requires exemption from the 2001 Directive as a whole. In 2002, the province of Upper Austria drafted a law prohibiting the cultivation of GM crops on its territory with the objectives of protecting GM-free production systems, the environment and natural biodiversity. The request cited a study in order to provide new scientific evidence. To argue that the province was uniquely affected, the government pointed out the high concentration of organic farmers in that province and the small-structured Austrian farming sector, which made the development of coexistence rules unfeasible. The ban would ensure that the production of organic farmers would not be contaminated and would also safeguard biodiversity.

The Commission dismissed the derogation request. It found that the study on which the Austrian claim was based while itself produced after the adoption of the 2001 Directive, was based on scientific evidence or information that was available earlier and thus constituted merely a validation of previous works rather than new evidence. It noted that 'Austrian concerns about coexistence relate more to a socio-economic problem than to the protection of the environment'[22] and this is not one of the grounds on which derogation requests can be made according to the criteria in Article 114(5) TFEU. Furthermore, it found that small-structured farming is not a feature specific to this region exclusively but is found in other Member States as well. Finally, the Commission noted that reliance on the precautionary principle is also misplaced in the Upper Austrian application (being one of a blanket ban on all GMOs) because 'a preventive measure cannot properly be based on a purely hypothetical approach to the risk, founded on mere conjecture, which has not yet been scientifically verified'.[23]

Some years later, Poland similarly defended its draft law establishing GM-free zones with, among other things, 'the need to fulfil the expectations of Polish society' and 'a high level of fragmentation of Polish agriculture'.[24] The Commission again rejected this application for a derogation on the grounds that the issues raised did not constitute new and additional scientific evidence and therefore did not fulfil the criteria needed to secure exemption.

In sum, Article 114(5) TFEU also allows an opt-out only on scientific grounds, to the exclusion of socio-economic and political ones. Moreover, as it is applied in practice, even the scientific conditions appear to be quite restrictive. The cumulative criteria of providing new scientific evidence and demonstrating the existence of conditions that uniquely affect the country, make it extremely difficult for a country to mount a successful case and in practice this mechanism has never secured an exemption for any Member State so far.

Evaluating the Functioning of the Accommodating Mechanisms

This section has examined the use of two modes of accommodation. Under the first mode, accommodation of different types of concerns should take place during the decision-making process. A second mode of accommodation involves the possibility of individual deviations from commonly adopted decisions. Under that mode, two specific mechanisms were reviewed: the use of a directive-specific safeguard clause and the derogation procedure based on Article 114(5) TFEU.

Several observations need to be highlighted. First, in some of the cases there is a mismatch between the formal framework and its application in practice with regard to accommodation of diversity. For example, the decision-making process provides some limited basis for the Commission to deviate from EFSA opinions and to go beyond scientific considerations, which the risk assessment part should be formally based on. However, in practice, the Commission has not done this so far. Conversely, the safeguard clause is intended as an emergency mechanism for situations where a rapid response is needed to prevent a given danger to health or the environment. It is not meant to function formally as an accommodating mechanism, taking a wide range of considerations. In practice, however, this mechanism is used to accommodate non-scientific concerns as bans have been preserved contrary to the EFSA opinions declaring those bans scientifically unfounded.

Focusing specifically on the actual application of the regulatory framework, the analysis suggests that both the decision-making process and Article 114(5) TFEU do not allow space for non-scientific concerns to affect the outcome. The safeguard clause, due to the involvement of political actors such as the national executives in the Council, produces more variegated outcomes and hence can be considered to perform better as an accommodating mechanism despite the fact that it was not designed to function as such.

The analysis of the types of accommodation formally allowed by the regulatory framework and their actual use demonstrate that the GMO authorisation regime privileges one type of knowledge over another (Chalmers 2003). It is predominantly based on science and therefore does not allow sufficient flexibility, given the divergent positions of Member States, to take into account the wider range of considerations rather than the exclusively scientific.

National Responses and the Way Forward for EU GMO Regulation

Given the above, it is perhaps not surprising that GMO authorisations experience chronic difficulties. In practice, it has mostly been the Commission that has taken the final decision because the Regulatory Committee and the Council have very rarely succeeded in adopting a decision, which stands in sharp contrast to the smooth functioning of the comitology procedure across other policy areas (Pollack and Shaffer 2010: 349). This also leads to delays and deadlocks, especially with regard to authorisations for cultivation purposes. For example, the authorisation of the GM crop, the Amflora potato, took 13 years from the time the application was lodged in 1996 to the issue of the official consent in 2010.[25] Thus, the functioning of the regulatory regime is far from satisfactory for the majority if not all of the actors involved (Brosset 2004). Biotech companies and pro-GMO Member States are frustrated with the tortuous authorisation procedure that can be strung out over several years and is wrought with uncertainties which make investment in the development of new GM varieties untenable.[26] Environmental organisations and consumer groups as well as GMO-sceptical Member States are dissatisfied with the eventual – even if delayed – approval of GMOs, critical of EFSA for always finding that GMOs present negligible risks to human and animal health and the environment, and the impossibility of working non-scientific considerations into the procedure. In addition, the number of Member States relying on the safeguard procedure with regards to GM maize and GM potato for cultivation, is growing. Finally, the Commission too is in an unenviable position, being under fire from all sides. It is under pressure to adopt decisions that in some cases go against the plurality of Member States but at the same time it needs to follow the formal procedure lest it be faced with lawsuits from the biotech industry or at international level (Skogstad 2011).

Recent developments point to ways for resolving the deadlock. Two directions for reform have been advocated by the Member States, which, in fact, correspond to the two modes of accommodation discussed above. The first reform initiative invites the Commission to take steps to open up the decision-making process at the EU level to a broader range of considerations. Under the leadership of France, which held the EU presidency in the second half of 2008, the Council unanimously called for strengthened environmental risk assessment, more freedom for Member States to decide on GM-free zones, and – most importantly for this analysis – appraisal of socio-economic benefits and risks.[27] The Council asked the Commission to begin to consider ways in which 'other legitimate factors' such as socio-economic considerations can be brought into the decision-making process. To that purpose, the Commission launched a consultation procedure soliciting input from the Member States on what they consider to fall within the scope of socio-economic aspects that should be taken into account when authorising GMOs. The national submissions would provide the basis for a report outlining a possible common definition of what 'socio-economic considerations' consist of as a first step toward working out a way to include non-scientific concerns in the regulatory framework. This direction of change would improve the functioning of the first mode of accommodation so that a wider range of concerns could be mediated through the decision-making process itself. The report was completed in April 2011 but it was concluded that the submissions in the Report did not provide enough common ground for the elaboration of a shared definition.[28]

A second line of reform takes the approach of strengthening possibilities for more differentiation. Rather than searching for ways to make common decision-making more inclusive and responsive to national diversities, the Dutch and Austrian delegations tabled proposals to reform the framework to allow Member States the freedom to ban the cultivation of GM crops authorised at the EU level. In the Environment Council of March 2009, the Netherlands came up with a Declaration proposing that Member States have the right to unilaterally decide on issues of GMO cultivation. This initiative was taken up and further developed by Austria in the subsequent Council meeting of June 2009.[29] It suggested the introduction of an opt-out clause that gives Member States the right to prohibit the cultivation of GMOs on their territory. The Austrian declaration was supported by 12 Member States. It is worth noting that both Austria and the Netherlands advocate what would in effect be the devolution of competences to the national level, yet for very different reasons because they stand on the opposite ends of the spectrum with regard to their position on GMOs. The Netherlands has traditionally supported genetic engineering and is frustrated that the EU regulatory process slows down its progress in this field (Somsen 2009). On the other hand, Austria – perhaps most sceptical towards GM technology in the EU – supports such devolution so that it can ban the cultivation of GMOs on its territory.

Both the French and joint Austrian–Dutch initiatives suggest ways in which the current inflexibility of the regime and the deadlock that it has created may be overcome. But underlying these two suggestions for reform are two very different approaches to the integration process. The proposal for incorporating broader sets of concerns into the decision-making process reveals a willingness to work together at the European level and to further integration by agreeing on a common understanding of what socio-economic factors mean. Alternatively, leaving the decision to the Member States amounts to rolling back competences to the national level and acknowledging the limits of the integration process in the spirit of 'we agree to disagree'.

The Commission has indicated a preference for the second line of reform. In his 'Political Guidelines for the next Commission' from the September 2009, Commission President Barroso stated:[30]

> I want to be rigorous about where we need to have common rules and where we need only a common framework. We have not always got the balance right, and we have not always thought through the consequences of diversity in an EU of 27. In an area like GMOs, for example, it should be possible to combine a Community authorisation system, based on science, with freedom for Member States to decide whether or not they wish to cultivate GM crops on their territory.

Following up on the guidelines of its President, in July 2010 the Commission tabled a reform package consisting of two parts.[31] It issued a new Commission Recommendation, effective immediately, on co-existence between GM and non-GM (conventional and organic) crops, giving more flexibility to Member States than the previous one. The second measure in the Commission's reform package is a proposal to amend the Deliberative Release Directive through the addition of an article (Article 26b in Directive 2001/18) that would allow Member States to restrict or prohibit the cultivation of GM crops on grounds other than those related to the protection of human or animal health and the environment. The amendment is subject to the co-decision procedure. The Commission proposal was initially received with many questions and reservations in the Council. This led the Commission to issue an indicative list of grounds for the restriction of prohibition of GM cultivation which seems to have pushed forward the adoption process.[32] Meanwhile, in July 2011, the proposal passed first reading in the European Parliament. The EP introduced an amendment to the text of the opt-out article, allowing for the use of environmental grounds and thus expanding the scope of the opt-out clause compared to what the Commission had originally proposed.[33] In March 2012, the Council failed to agree on a compromise text prepared under the Danish Presidency. At the time of writing, the fate of the proposal is still not sealed and the exact form of the opt-out clause is subject to change. Yet, it is clear that the reform focusing on accommodation through differentiation is at a more developed stage and has higher chances to yield a result in the near future.

Conclusion

The analysis above has developed the argument that accommodation of national diversity in the area of GMO authorisations is mainly based on science. Non-scientific claims for accommodation, such as socio-economic concerns, preferences for certain types of agriculture, and public opinion, are largely excluded from the decision-making process. Part of the reason is that the Commission underutilises its role as risk manager in the regulatory stage of the authorisation process. Instead of working out mechanisms and procedures through which Member States can channel the broad range of concerns that determine their positions, the Commission tries to sideline the contestation of opposing views by appealing to the principles of science-based policy making. But rather than depoliticising the procedure as a whole, this simply shifts the turf on which the conflict takes place to the scientific realm. It places EFSA at the focal point of the contestation and transforms the conflict into one between EU science and national science.

In part, the stance of the Commission is understandable due to its external and internal commitments. Externally, the Commission acts in the shadow of the WTO dispute. Moreover, with regard to its internal obligations, the Commission is responsible for safeguarding the integrity and the seamless functioning of the Internal Market and, as such, it looks unfavourably on the segmentation of the EU market into national markets. Yet, to a great extent the reason why the Commission grabs on to the appeal of scientific universalism is because as a technocratic body it simply does not have the mechanisms at its disposal to resolve such politically charged issues.

A further observation following on from the preceding analysis is the mismatch between the formal framework and the way it is implemented in practice. On the one hand, the EU has in place a regulatory procedure (in its risk management part) that, as it formally recognises, has to tackle political tasks, and yet ends up being based solely on science. On the other hand, a legal mechanism such as a safeguard clause that is formally and exclusively based on science and is meant to serve as an emergency button in cases of unforeseen problems in the food chain or in the environment is, in practice, used for political ends and allowing in some limited cases a degree of accommodation.

The reform currently underway aims to remedy this bias by introducing an opt-out clause to allow the accommodation of non-scientific claims. This would also obviate the need to use the safeguard clause as an accommodating mechanism and thus eliminate some of the mismatch between the formal framework and its functioning in practice.

It is telling that between the two reform options – including non-scientific concerns in the decision-making process and allowing differentiation in the post-decision-making phase – the Commission preferred the second. It underscores that in the context of a salient issue the Commission prefers to give Member States the freedom to decide for themselves rather than try to work out a solution at the EU level.

The reform proposed by the Commission is remarkable also because it essentially suggests devolving power back to the Member States, in effect rolling back competences to the national level. Whereas before functionally equivalent opt-out measures were sanctioned at the EU level, with the introduction of the new Article, this will be unilaterally done by Member States.

The implications of the preceding analysis point to ways that policy making has the potential to affect constitutional characteristics of the EU, such as the division of competences between the Member States and the EU. It demonstrates how features of the regulatory framework – and specifically its lack of flexibility towards the accommodation of non-scientific national concerns – can precipitate reforms that redistribute political authority among various levels of governance.

Notes

1 I am greatly indebted to Prof. Peter A. Hall for generously offering many valuable suggestions for developing the arguments in this piece. I would also like to thank my interview partners for sharing their time, expertise and insights with me as well as the editors of this volume for their helpful comments. I gratefully acknowledge the financial support for this research provided by the Minda de Gunzburg Center for European Studies at Harvard University. An earlier version of this chapter was presented at the Workshop for Young Researchers, 'European Integration between Trade and Non Trade: Selected Issues' at the University of Maastricht in April 2010.

2 See Pollack and Shaffer (2010) for a more detailed general overview of the developments in the policy area.

3 Council Directive 90/219 on the contained use of genetically modified micro-organisms, OJ 1990 L 117/1.

4 Council Directive 90/220 on the deliberate release into the environment of genetically modified organisms, OJ 1990 L 117/15.

5 COM (1999) 719 final of 12 January 1999.

6 Regulation 178/2002 of the European Parliament and of the Council of 28 January 2002 laying down the general principles and requirements of food law, establishing the European Food Safety Authority and laying down procedures in matters of food safety, OJ 2002 L 31/1.

7 Before the reform, EFSA's functions were performed by committees of national experts managed by the Commission.

8 The comitology procedure involves subject-specific committees of national representatives, voting to approve or reject implementing acts proposed by the Commission. The post-Lisbon comitology procedure is guided by Regulation 182/2011 of the European Parliament and of the Council of 16 February 2011 laying down the rules and general principles concerning mechanisms for control by Member States of the Commission's exercise of implementing power, OJ 2011 L 55/13.

9 ECJ, Cases C-157/96 and C-180/96 of 5 May 1998.

10 Art. 7, Regulation 178/2002, OJ 2002 L 31/1.

11 COM(2000) 1 final of 2 February 2000.

12 Regulation 1830/2003 of the European Parliament and of the Council of 22 September 2003 concerning the traceability and labelling of genetically modified organisms and the traceability of food and feed products produced from genetically modified organisms and amending Directive 2001/18, OJ 2003 L 268/24.

13 Commission Recommendation of 13 July 2010 on guidelines for the development of national co-existence measures to avoid the unintended presence of GMOs in conventional and organic crops, OJ 2010 C 2001, as repealed pursuant to Commission Recommendation 2003/556 of 23 July 2003 on guidelines for the development of national strategies and best practices to ensure the coexistence of genetically modified crops with conventional and organic farming, OJ 2003 L 189/36.

14 A process-oriented approach is based on the premise that products obtained through a novel production process (e.g. genetic engineering) warrant separate and additional regulation. This is the approach preferred by the EU. According to a product-oriented approach, regulation is not necessary if the final product is deemed equivalent to its conventional counterpart, regardless of the process through which it is obtained. This approach is employed by the US and other GM-producing countries. These two regulatory approaches came into conflict within the free trade framework of the WTO. See *European Communities – Measures Affecting the Approval and Marketing of Biotech Products*, Reports of the Panel, WTO Docs WT/DS291/R, WT/DS292/R, WT/DS293/R of 29 September 2006.

15 Pollack and Shaffer (2009: 260) refer to the overall revision of the regulatory framework as 'reform without change'. Brosset (2004: 555) notes, '[…]some changes to the procedure were implemented 2001 but they were not intended to restructure the general architecture of the procedure established in 1990'.

16 The authorisation procedure under Regulation 1829/2003 is slightly different and more centralised. Under it, the competent national authority with which the application is lodged does not perform the initial risk assessment. This task is directly transferred to EFSA, followed by a draft decision proposed by the Commission and adopted through the comitology procedure described above. In practice, the two tracks are almost identical because initial national risk assessments under the 2001 Directive are typically challenged by an objecting Member State and the risk assessment is pushed up to the EU level from where on the two processes run in the same way.

17 Similarly, Chalmers (2003: 560) puts it in terms of a 'process of mutual adaptation' or a 'principle of alignment' whereby there is the 'duty to listen to interests adversely affected by a decision' and 'a duty to show that other values have been considered'.

18 The safeguard clause in Regulation 1829/2003 makes direct reference to the emergency clauses in the General Food Law (Arts. 53 and 54 of Regulation 178/2002). The conditions according to which emergency measures can be initiated are again based on scientific evidence such as the likelihood of 'serious risk to human health, animal health, or the environment'. Worthy of note here is a recent judgment from September 2011 in which the Court of Justice of the European Union determined that a French safeguard measure banning the GM maize MON810 was improperly based on Art. 23 of the Deliberate Release Directive and should have instead been based on Art. 34 of the 2003 Regulation (ECJ, Joined Cases C-58/10 to C-68/10). It remains to be seen whether the French authorities will take steps to invoke the ban under Art. 34 of the 2003 Regulation and to what effect.

19 Some of the more recently placed bans have not been challenged yet by Commission decision proposing to lift the bans because EFSA opinion is still pending.

20 Council of the European Union. 2670th Council Meeting, 24 June 2005, 10074/05, Presse 174.

21 Commission Decision 2003/653, OJ 2003 L 230/34; Commission Decision 2008/62 OJ 2008 L 16/17; Notification pursuant to Art. 95(5) EC Treaty, OJ 2009 C 139/2.

22 Commission Decision 2003/653, OJ 2003 L 230/34, at p. 42.

23 Commission Decision 2003/653, OJ 2003 L 230/34, at p. 43.

24 Commission Decision 2008/62, OJ 2008 L 16/17, at p. 24.

25 BASF Press Release from 2 March 2010 'Commission approves Amflora starch potato', available at: <http://www.basf.com/group/pressrelease/P-10-179> (last accessed on 17 November 2012).
26 In fact BASF, the company that developed one of the two GM crops currently authorised for cultivation in the EU, announced its plans to stop working on developing GM varieties for the European market and relocate to the United States in January 2012 ('An End to GM Crop Development in Europe', *Financial Times*, 16 January 2012).
27 Council of the European Union, 16882/08 of 4 December 2008 at p. 5.
28 Report from the Commission to the European Parliament and the Council on socio-economic implications of GMO cultivation on the basis of Member States contributions, as requested by the Conclusions of the Environment Council of December 2008, available at: <http://ec.europa.eu/food/food/biotechnology/reports_studies/docs/socio_economic_report_GMO_en.pdf> (last accessed on 17 November 2012).
29 Council of the European Union, 11226/2/09 of 24 June 2009.
30 Available at: <http://ec.europa.eu/commission_2010–2014/president/pdf/press_2009 0903_en.pdf>, p. 39 (last accessed on 17 November 2012).
31 Communication from the Commission to the European Parliament, the Council, the Economic and Social Committee and the Committee of the Regions on the *Freedom for Member States to decide on the cultivation of genetically modified crops*, COM(2010) 375 final of 13 July 2010.
32 SEC (2011) 184 final of 8 February 2011.
33 'GMOs: Parliament backs national right to cultivation bans', available at: <http://www.europarl.europa.eu/news/en/headlines/content/20110627FCS22686/8/html/GMOs-Parliament-backs-national-right-to-cultivation-bans> (last accessed on 17 November 2012).

Bibliography

Barroso, J.M. (2009) 'Political Guidelines for the Next Commission', available online: <http://ec.europa.eu/commission_2010-2014/president/pdf/press_20090903_en.pdf> (accessed 21 September 2012).

Brosset, E. (2004) 'The Prior Authorisation Procedure Adopted for the Deliberate Release into the Environment of Genetically Modified Organisms: The Complexities of Balancing Community and National Competences', *European Law Journal*, 10(5): 555–79.

Chalmers, D. (2003) '"Food for Thought" Reconciling European Risks and Traditional Ways of Life', *Modern Law Review*, 66(4): 532–62.

——(2005) 'Risk, Anxiety and the Mediation of the Politics of Life', *European Law Review*, 30: 649–74.

Everson, M. and Vos E. (2009) 'The Scientification of Politics and the Politicisation of Science', in M. Everson and E. Vos (eds), *Uncertain Risks Regulated*, New York: Routledge-Cavendish, 1–17.

Fleurke, F. (2008) 'What Use for Article 95(5) EC?', *Journal of Environmental Law*, 20(2): 267–78.

Joerges, C. (2001) 'Law, Science and the Management of Risks to Health at the National, European and International Level – Stories on Baby Dummies, Mad Cows and Hormones in Beef', *Columbia Journal of European Law*, 7: 1–19.

Kritikos, M. (2009) 'Traditional Risk Analysis and Releases of GMOs into the European Union: Space for Non-scientific Factors?', *European Law Review*, 34(3): 405–32.

Poli, S. (2004) 'The Overhaul of the European Legislation on GMOs, Genetically Modified Food and Feed: Mission Accomplished. What Now?', *Maastricht Journal of European and Comparative Law*, 11(1): 13–45.

Pollack, M. and Shaffer, G. (2010) 'Biotechnology Policy', in H. Wallace, M. Pollack and A. Young (eds), *Policy-Making in the European Union*, New York: Oxford University Press, 331–55.

Pollack, M. and Shaffer, G. (2009) *When Cooperation Fails*. New York: Oxford University Press.

Skogstad, G. (2011) 'Contested Accountability Claims and GMO Regulation in the European Union', *Journal of Common Market Studies*, 49: 895–915.

Somsen, H. (2009) 'GMO Regulation in Netherlands: A Story of Hope, Fear and the Limits of 'Poldering'', in M. Everson and E. Vos (eds), *Uncertain Risks Regulated*, New York: Routledge-Cavendish, 187–206.

6 The Politics of Risk Decision Making

The Voting Behaviour of the EU Member States on GMOs

Madjar Navah, Esther Versluis and Marjolein B.A. van Asselt

Introduction

Genetically modified organisms (GMOs) have been prevalent in the European Union (EU) for nearly two decades, mainly through import. In the context of agro-biotechnology, GMOs are designed, for example, to be more resistant to diseases, tastier or faster growing, in order to increase the economic value of agricultural products. The topic of GMOs in the EU has been an international issue from the outset. Decision-making on GMOs, both at the EU and the Member State level, has largely been about deciding on the acceptance of imported agricultural products – whether end products or seeds – or the wish of companies based outside the EU to grow GMOs in Europe.

As is common with the introduction of new technologies on the market (Nowotny *et al.* 2001), public and academic discussions as well as scientific uncertainty have accompanied the introduction of GMOs across the European Union (e.g. Hunter 1999; Levidov 2001; Marchant 2001; Tulloch and Lupton 2002; Gaskell *et al.* 2004; Löfstedt *et al.* 2004; Vos and Wendler 2006; Van Asselt and Vos 2008). Induced by the BSE crisis of the mid-1990s, citizens in several European countries lost their trust in the agricultural food sector (Eurobarometer 2006). Particularly in recent years, backed up by green political as well as social movements, the contention surrounding GM-food has risen significantly. Opponents of GMOs are active in Brussels and the national capitals as well as on the Internet, lobbying for their cause and pointing their fingers at agro-biotechnological companies such as Monsanto and BASF (FOE-Europe 2011; Greenpeace 2011a, 2011b). This has recently cumulated in a citizen's petition on GMOs – made possible since the ratification of the Lisbon Treaty – that has been successfully handed over to the European Commission (Commission 2010). In addition to these societal concerns, there has been growing academic interest in the scientific uncertainty relating to GMOs. The effects of GM products on the environment as well as human and animal health have not yet been determined conclusively (e.g. Van Asselt and Vos 2006; see also Hristova in this volume).

In sum, GMOs lead to regulatory controversy in the context of uncertainty and free trade. Deciding about GMOs entails political decision-making designed to balance between trade and risk (see Zurek in this volume for more on GMOs and trade). While the question of how the EU authorities come to their political decisions has been covered (see amongst others Hristova in this volume), we have less insight into political decision-making on GMOs from an angle that does not solely discuss normative or scientific concerns surrounding this technology, but rather accounts for the political processes. Awareness of the issue of genetically modified food amongst citizens and public media are prevalent but this attention has evolved rather and has concentrated on the risks and normative opinions on the topic (Eurobarometer 2006) as well as on the introduction of individual GM crops (e.g. for Germany: Agrarheute 2011). Interest in the political dimension has been marginal.

One way to examine the political sphere is to investigate voting behaviour. Voting is critical in the political process, as it is a key principle of democratic decision-making and it is the moment in which politicians have to take sides and show their colours. So far, an overview of how the 27 Member States of the EU have voted on various GMOs at the European level is lacking. Jasanoff's 'Designs on Nature' (2007) – comparing the politics and policies on biotechnology in the US, Germany, the UK and the EU – convincingly demonstrated that political diversity between countries matters in decision-making on GMOs. Jasanoff examined the policy divergences and the legitimacy of state actions. By analysing concepts of democratic theory within the politics of science and technology, Jasanoff explained that 'nation-building' projects and the political culture in the countries are at stake. Likewise, the chapter by Hristova in this volume demonstrated divergence between the EU Member States when it comes to GMO regulation, and described the importance for the EU of being able to accommodate such diversity.

This chapter contributes to the current debate on the regulation of trade on GMOs in the EU by providing insight into the importance of domestic politics for the EU decision-making process. By empirically evaluating 18 decision-making procedures in the Standing Committee on Food Chain and Animal Health (SCFCAH) and in the Council of the EU, we analyse the (domestic) political rationale behind EU decision-making on GMOs. This will encourage a better understanding of what types of risk and trade concerns in Member States are translated into voting positions. Perhaps this analysis will help us to better understand why the Member States have such difficulty in reaching a qualified majority when deciding on the approval for the placing of specific GMOs on the Internal Market, while they more easily established consensus when discussing Commission proposals to lift national safeguard bans (see Hristova in this volume). We advance a comparative political approach: we first provide insights into the 27 different national views, which are relevant if we want to delve into the next layer of understanding of the political discussion on GMOs in the EU. We question whether any patterns in Member States' voting can be identified. Can we identify States that always vote in favour of GMOs? Do other States generally vote against GMOs? Is the pattern more dispersed and do we see countries switching position from one vote to another? What do such patterns (or the lack of patterns) teach us about the politics of

GMO decision-making? How can we account for these patterns? Do changes in government at Member State level translate into changes in voting behaviour at the European Union level? Or is it possible to trace voting positions to political party affiliations such as the right or the left?

This chapter starts by outlining GMO legislation in the EU and describing the GMO authorisation procedure. Thereafter, we analyse past decision-making procedures of both the SCFCAH and the Council of the EU on GMO cases, and categorise countries according to their voting behaviour. The chapter then searches for patterns in party lines in the respective Member States taking a similar stance on GMOs before questioning whether a change in government has any effect on a country's position on biotechnology. To conclude, we will look behind the curtain of this much contested issue and highlight the importance of identifying the domestic political rationale in order to understand the decision-making on risk–trade issues.

Genetically Modified Organisms, European Legislation and the Long Road to Authorisation

Before being able to understand voting behaviour, we first need to sketch the context by describing the political process at the EU level. At what stage, and in what form, do the Member States play a role in the most important procedure relating to GMOs: the authorisation procedure?

As defined in Article 2 of the Directive 2001/18 on Deliberate Release into the Environment of Genetically Modified Organisms, an organism is considered as modified genetically if the DNA has been changed in such a manner that could not have been achieved by aid of cross-breeding or natural recombination in natural conditions. Since its first confrontation with GMOs in the early 1990s, EU legislation has been enacted both to protect health and the environment while at the same time ensuring the unhindered movement of safe and healthy genetically altered products in the EU. So the regulation of GMOs has always been part of the Internal Market and the trade sphere.

The European Union foresees an explicit authorisation for food and feed made from GMOs in order to allow their entrance onto the European market. It is a decision as to whether or not GM feed and food can be traded with and within the EU. The ideal scenario would be that such an authorisation procedure is executed by the EU with the consequent conclusion being applicable to and applied in all Member States of the EU.[1] The EU regulation on genetically modified food and feed, Regulation 1829/2003, provides the basis for the three phases of the GMO-authorisation process. The most important actors in the procedure are the European Commission – responsible for both food and feed safety and biotechnology as well as health and consumer protection in the context of biotechnology labelling – and the European Food Safety Authority (EFSA) with its GMO Panel consisting of independent academics from different Member States but not representing them in political terms.

The authorisation procedure begins with the submission of the import or application of GMOs to the national authority of one of the 27 Member States where the respective GMO is to be marketed. This application must include academic reports stating that the GM food is neither detrimental to health nor the environment, showing that the GM food is equal to conventional equivalents for instance by analysis of certain components/ingredients, making recommendations for the labelling of products and providing the means and sample material in order to detect the GM substance. The submission may well take account of a post-market screening system. Furthermore, a summary of the whole submission dossier shall be included. Only if these requirements are fulfilled will the respective national authority convey the application to EFSA. EFSA subsequently informs all other Member States, thereby permitting them to see the application, and makes the summary of the submission dossier accessible to the public (GMO Compass 2011).

The second step in the authorisation process deals with the risk assessment. Given that all requirements and documents are at hand, EFSA formulates an opinion within six months as to whether or not the GMO is considered safe. It is possible to extend the period for reaching a decision should additional documents be demanded. As already indicated above, an expert panel on genetic engineering (GMO Panel) provides the academic evaluation underpinning the EFSA opinion by assessing whether the applicant's product remains within the margins of variability, which is found in its natural counterparts. Should a product require approval under Directive 2001/18 on the Deliberate Release into the Environment of Genetically Modified Organisms, the applicant has to prove that all actions avoiding detrimental effects on human health, flora and fauna have been taken. Advice from the Members States' national and federal authorities has to be sought. A further step for EFSA is to request the EU reference laboratory to assess and confirm the detection methods referred to in the submission documents. The official opinion also includes a proposal for the labelling of the product and an environmental monitoring scheme for the concerned seed. In addition, a framework might be established to respond quickly to outcomes of future monitoring. Finally, by forwarding its opinion to the Member States and the European Commission, the authority publishes the documents openly for European citizens[2] (GMO Compass 2011).

The final phase centres on the political decision. After having submitted the opinion to the Commission, the latter is required to draft a decision as to whether or not the GMO is to be approved for trade with and within the EU within 90 days. In case of deviations from EFSA's opinion, a written reasoning has to accompany the Commission's draft decision. The Commission forwards its draft decision to the Standing Committee on the Food Chain and Animal Health (SCFCAH), in which delegates from all 27 Member States are asked to approve or decline the proposed decision by qualified majority. This is the political arena. If this Committee differs in opinion with the Commission's draft or if a qualified majority decision is unobtainable, the Commission is required to call the Council of Ministers of the European Union and present its stance as well as to notify the European Parliament.

The Council of Ministers should approve or reject by qualified majority[3] the draft decision on import or application of the concerned GMO within the EU Internal Market within three months. Here, three scenarios are possible: 1) the Council explicitly approves the Commission's draft decision, implying that the decision is implemented by the Commission; 2) the Council disapproves the Commission's draft and the Commission has to amend or withdraw its draft, or; 3) the Council cannot arrive at a qualified majority for or against market authorisation, in which case the matter is referred back to the Commission, *de facto* making the Commission responsible for the final decision.[4] Hristova (in this volume; see also Van Asselt and Vos 2008) outlined that when the Member States are unable to reach a qualified majority in the Standing Committee and in the Council, the Commission always approves authorisation for the Internal Market. All authorisations made by the European Union in GMO cases are valid for ten years and authorised products are listed into a public catalogue.

We now analyse cases where the Standing Committee on Food and Council (only in particular subtexts if no decision has been reached at committee level) had to decide on draft Commission decisions. It is to be expected that each single decision was much contested with substantial but minority blocks of Member States 'in favour' and 'against' the Commission's proposal in the committee. If Member State delegates stick to pre-defined national positions, it is likely that the blocks present in the Committee on Food Chain will be reproduced in the Council.

Decision-making on GMOs in the European Union

Analysing different Council decisions and decisions taken by the SCFCAH proves to be a difficult task, with information sources not always being openly accessible. We made use of official documents issued by the European Union, ranging from minutes of meetings to Council decisions. Where information gaps were prevalent, the official EU citizen help centre was called for additional data and to synchronise and confirm the already collected data. In order to fill blind spots in the voting behaviours within the Council and the SCFCAH, NGOs, EU Research Programmes and independent information platforms – such as the GMO Compass – were also consulted/examined. Although the EU is not responsible for the content of the GMO Compass, the Commission financially supported the website initiative. The GMO Compass offers consumers not only information on the benefits and risks of GMOs but also aims to promote an informed debate by providing an extensive database on every GMO authorisation procedure, relevant safety assessments, planting sites and legal background (Co-Extra 2010).

Due to data collection problems both in terms of data availability and data quality, we only succeeded in fully tracing the voting behaviour of 18 voting procedures involving 11 different crops in the period 2003–5. We focussed on these years due to the relative availability of data. This number constitutes about one fifth of the already authorised crops (GMO Compass 2011). These crops are all maize crops for food and feed (with the exception of the rapeseed crop GT73), that contain modified improvements such as herbicides tolerance or insect resistance. The Swiss company Syngenta as well as the US companies Monsanto and Pioneers/Dow AgroSciences were the producers of the analysed GM crops. On these 18 procedures we succeeded in mapping out the respective voting behaviours of the Members States in the SCFCAH and the Council, enabling us to draft the comprehensive overview that had been lacking. Table 6.1 provides a full overview of 18 different decisions made by Member States and their respective delegates in either the SCFCAH or the Council of Ministers. Whether a vote was taken in the Council of Ministers is indicated

Table 6.1 Overview of 18 different decisions made by Member States and their respective delegates in either the SCFCAH or the Council of Ministers

	24 October 2005*	24 October 2005*	20 September 2005*	19 Sept. 2005 (StCo)	24 June 2005*
	GA 21	MON 863 (old novel food)	Dow/Pionner's maize 1507 (for animal feed)	MON 863 X MON810	MON863 (for feed)
Austria	X	X	X	X	X
Belgium	+	I	+	+	o
Cyprus	X	X	X	X	X
Czech R.	I	+	+	o	o
Denmark	X	X	+	X	X
Estonia	+	+	I	+	+
Finland	+	I	+	+	+
France	X	+	+	+	+
Germany	o	+	o	o	+
Greece	X	X	X	o	X
Hungary	o	o	X	X	X
Ireland	+	+	o	o	o
Italy	X	+	X	X	X
Latvia	X	X	o	X	X
Lithuania	X	X	X	X	X
Lux	X	X	X	X	X
Malta	X	X	X	X	X
NL	+	+	+	+	+
Poland	X	o	X	o	X
Portugal	X	X	X	X	X
Slovakia	X	X	o	X	X
Slovenia	o	X	o	X	X
Spain	o	o	o	o	o
Sweden	+	+	+	+	+
UK	+	+	+	+	+
TOTAL No (X)	155	74	120	103	142
TOTAL Yes (+)	94	181	123	104	121
TOTAL Abst. (o)	72	66	78	114	58
Total votes	321	321	321	321	321
QMV	232	232	232	232	232

Table 6.1 (continued)

	3 June 2005 (StCo)	17 May 2005 (StCo)	17 May 2005 (StCo)	27 April 2005 (StCo)	20 Dec. 2004*	29 Nov. 2004 (StCo)
	Dow/Pionner's maize 1507 (for food)	Dow/Pionner's maize 1507 (for animal feed)	MON863	GA 21	GT73	MON863 (for feed)
Austria	X	X	X	X	X	X
Belgium	X	+	+	+	X	o
Cyprus	X	X	absent	o	X	X
Czech R.	+	+	+	+	o	o
Denmark	o	o	X	o	X	o
Estonia	+	+	X	o	X	+
Finland	+	+	+	+	+	+
France	+	+	+	X	+	+
Germany	o	o	+	o	o	+
Greece	X	X	o	absent	X	X
Hungary	o	X	X	o	X	X
Ireland	+	o	o	+	o	o
Italy	o	X	X	o	X	X
Latvia	X	o	o	+	X	X
Lithuania	X	X	o	absent	X	X
Lux	X	X	o	X	X	X
Malta	o	o	o	absent	X	X
NL	+	+	+	+	+	+
Poland	o	X	+	o	X	X
Portugal	X	X	o	X	+	+
Slovakia	o	o	X	X	+	o
Slovenia	o	o	X	o	o	X
Spain	o	o	X	o	o	o
Sweden	+	+	+	+	+	+
UK	+	+	+	+	o	+
TOTAL No (X)	65	117	100	62	135	116
TOTAL Yes (+)	111	116	168	94	78	133
TOTAL Abst. (o) (including absent votes)	145	88	53	165	108	72
Total votes	321	321	321	321	321	321
QMV	232	232	232	232	232	232

	20 Sept. 2004 (ind.vote) (StCo)	09 July 2004*	28 June 2004*	16 June 2004 (StCo)	30 April 2004 (StCo)	26 April 2004*	08 Dec. 2003 (StCo)
	MON863	NK603 (for human food)	NK603 (for animal feed)	GT73	NK603 (for human food)	Bt11	Bt11
Austria	X	X	X	X	X	X	X
Belgium	X	+	o	+	+	o	o
Denmark	X	X	X	X	X	X	X
Finland	+	+	+	+	+	+	+
France	X	+	+	+	+	X	X
Germany	o	o	o	o	o	o	o

Table 6.1 (continued)

	20 Sept. 2004 (ind.vote) (StCo)	09 July 2004*	28 June 2004*	16 June 2004 (StCo)	30 April 2004 (StCo)	26 April 2004*	08 Dec. 2003 (StCo)
	MON863	NK603 (for human food)	NK603 (for animal feed)	GT73	NK603 (for human food)	Bt11	Bt11
Greece	X	X	X	X	X	X	X
Ireland	o	+	+	o	+	+	+
Italy	X	X	X	X	+	+	o
Lux	X	X	X	X	X	X	X
NL	+	+	+	+	+	+	+
Portugal	o	+	+	+	X	X	X
Spain	o	o	o	o	o	o	+
Sweden	+	+	+	+	+	+	+
UK	+	+	+	X	+	+	+
New MS							
Cyprus	X	X	X	X			
Czech R.	o	+	+	+			
Estonia	+	o	+	X			
Hungary	X	o	X	X			
Latvia	X	I	X	+			
Lithuania	X	X	X	X			
Malta	X	X	X	X			
Poland	X	o	+	X			
Slovakia	o	o	+	+			
Slovenia	X	o	o	o			
TOTAL No (X)	65	31	39	57	19	29	29
TOTAL Yes (+)	21	53	59	43	50	35	33
TOTAL Abst. (o)	34	40	26	24	18	23	25
Total votes	124	124	124	124	87	87	87
QMV	88	88	88	88	62	62	62

Source: StCo = vote Standing Committee; * = vote Council of Ministers. Compiled from the following sources: Biosafety Information Centre, 2004; EurLex, 2004; Europa, 2004; FOE-Europe, 2010; FOE-UK, 2004; ISIS, 2004; SOS, 2004).

by an asterisk while the abbreviation '(StCo)' indicates that the vote was taken in the SCFCAH.

To enhance analysis, the data have also been translated into a 'histogram' for the EU-15 and the EU-10 Member States in Figures 6.1 and 6.2. In these figures voting behaviour is clustered per country, which enables us to see whether there is a consistent pro- or contra-GMO attitude in their voting behaviour over the 3-year period. Austria is the only country with a 100 per cent anti-GMO voting pattern, while Cyprus, Greece, Hungary, Lithuania, Luxemburg, Malta and Slovenia never voted in favour of an approval between 2003 and 2005. On the other hand, between 2003 and 2005 the Netherlands, Sweden and Finland consistently voted in favour of GMO approval, while Germany, Ireland and the Czech Republic never voted against. It is also remarkable that Spain and Germany abstained or were absent in the voting in 90 per cent and 80 per cent of the voting

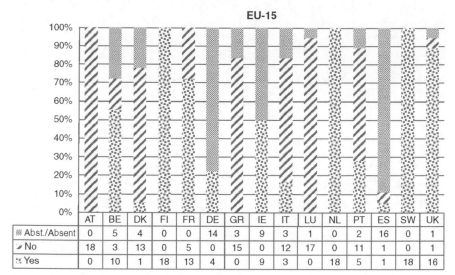

Figure 6.1 EU-15

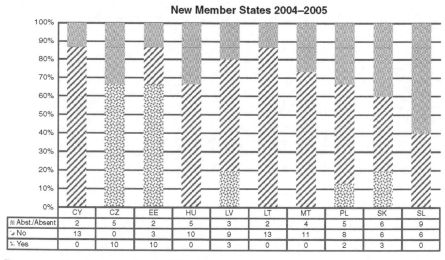

Figure 6.2 New Member States 2004–5

procedures respectively. This political diversity contrasts with the portrayal of Europe in the seminal analysis by Jasanoff (2007) of a prevalent political culture in European GMO dealings.

In none of the 18 cases was a qualified majority reached between the Member States, during the first votes in the SCFCAH nor when the Council convened in order to take a decision. It is noteworthy that the Council is 'only' asked to decide on the authorisation of a GMO crop if the standing committee was

not able to reach a qualified majority. This provision for extraordinary circumstances is also apparent in the standard practice: the Council had to
convene often because reaching a majority is difficult at the standing committee level. As expected, we hardly witnessed any differences in Member
State voting behaviour between the committee and Council levels. The
Member States did not, however, manage to form a majority to block the
authorisation of each new GMO. Consequently, the application, as outlined
before, was referred back to the Commission, giving it the final say in all 11
procedures, and in fact it authorised all GM crops.

As the Commission is a technocratic body and not a political forum, this
means that in practice in these cases there was not just a democratic deficit
but also a political deficit (the country representatives participating in the standing committee and Council voting procedures are not elected parliamentarians
and in many cases the accountability relationship between the country
representatives and the parliament at the Member State level is weak)
(Follesdal and Hix 2005; Majone 2005; Van Asselt and Vos 2008; see also
Borrás 2006): the decision is not taken in a political forum.

The lack of qualified majority pro or contra reflects maximum political
diversity and controversy. Against the background of the societal concerns,
it is not so surprising that the technocratic approval in situations where the
political qualified majority about contested products is lacking gave rise to the
situation where the Member States decided to invoke EU safeguard clauses in
order to ban certain GMOs (see also Ivanova and Van Asselt in this volume).

If we compare the data we compiled and discussed above with more
recent accounts on the Member States' stances on GMOs provided in the
Global Agricultural Information Network published by the United States
Department of Agriculture 'GAIN Report 2009', the following clusters of
Member States may roughly be identified:

- Negative on GMOs: Austria, Cyprus, Denmark, Greece, Hungary, Lithuania,
  Luxembourg and Malta; Poland and Portugal (since 2005); Ireland (more
  recently).
- Positive on GMOs: Czech Republic, Estonia, Finland, the Netherlands,
  Spain,[5] Sweden and the United Kingdom; Ireland (2003–5).
- Mixed view on GMOs: Belgium, France, Italy and Latvia; Germany (can
  also be seen as slightly critical, although without actively having voted
  against the introduction of GMOs); Poland (2003–5); Portugal (before 2005).

When comparing our data with stances outlined in the GAIN Report, three
countries expressed a significant change in their stance on GMOs within six
years: Ireland, Poland and Portugal. Ireland changed from pro-GMO between
2003 and 2005 to anti-GMO in 2009. In 2003–5, Poland and Portugal
demonstrated a mixed view, which changed into an anti-GMO stance from
2005 onwards. So over time, all three became more negative in their views.

However, we have no evidence for a change from a negative or mixed to a
positive attitude in any of the Member States. This pattern in changes in

political views contradicts the view that over time a new technology and associated products become less contested as it becomes more familiar and benefits pay-off, unless potential risks manifest themselves quite heavily or other social amplification factors (Pidgeon *et al.* 2003) increase to apply.

Assessing the Results: Looking Behind the Curtains of Political Decisions

How can we explain the identified patterns in voting behaviours? As the bulk of the necessary minutes, statements and details on decisions is not publicly accessible, we will not be able to further analyse Member States' policies on each single GM authorisation procedure. We have no sources to examine in more detail why 'mixed' countries voted differently in different cases or on what grounds certain Member States are clearly pro or contra the trade of GM products on the European market. In order to better understand this political behaviour, we decided to examine the governments in power at the Member State level and their political affiliations in terms of the classic right-centre-left typology. We hypothesised that this might enable us to better understand or at least characterise a country's stance on GMOs. Comparison with government affiliations would at least allow us to explore whether or not views differ along ideological lines.

In Table 6.2 we summarised our findings pertaining to the government in power and political ideology. Ireland, Poland and Portugal have changed their position on GMOs within the last six years. Can these changes be understood in terms of a change in government in these countries? Table 6.2 indicates that each of these three countries has indeed experienced a change in government. However, other countries also experienced changes in government. What can we conclude from comparison between the patterns in GMO voting and the analysis of governments at the Member State level?

Green Anti-GMO Countries?

The EU Member States that are almost consistently anti-GMO are Denmark, Greece, Luxembourg, Lithuania, Austria, Poland (since 2005), Hungary, Cyprus, Ireland (since 2007), Malta and Portugal (since 2005). Many of these countries[6] issued bans at the Member State level with reference to the safeguard clause. Furthermore, Greek, Austrian and Polish regions have recently joined the alliance of GMO-free regions in Europe, a formal network, whilst Ireland's green government party aims at being a GM-free island (GAIN Report 2009; GMO-free Europe 2010). Although having a rather mixed record of voting on GMO cases in the case of France or even having never voted against in the case of Germany, France (after the Sarkozy-government came to power in 2007) and Germany (until the 2009 elections when the Liberals entered the government coalition) also issued several bans

Table 6.2 Selected Member States, their voting behaviour on GMOs and their governments' political affiliation

Member State	Pro/ Contra GMOs 2003-5/today	Government's political affiliation (as in 2004/5)	Government's political affiliation (as in 2010)
Austria	Contra	Centre-Right/ Far-Right	Social-Democrat/ Centre-Right
Czech Republic	Pro	Centre-Right/ Social-Democrat/ Liberal	Centre-Right/ Centrist/Green
Denmark	Contra	Liberal/Conservative	Liberal/Conservative
Finland	Mixed/Pro, but stricter	Centrist/ Social-Democrat	Centrist/Centre-Right/Green
Germany	Mixed/Slightly Pro	Social-Democrat/ Green	Liberal/Conservative
Greece	Contra	Centre-Right	Socialist
Ireland	Pro/Contra	Centre-Right/ Centrist	Centrist-Green
Poland	Mixed/Contra	Centre-Left	Centre-Right
Portugal	Mixed/Contra	Social-Democrat/ Right-Wing	Centre-Left
Romania	Contra	Social-Democrat	Centre-Right
Spain	Pro	Centre-Right	Centre-Left
Sweden	Pro	Social-Democrat	Centrist/Liberal
The Netherlands	Pro	Centre-Left/Centre-Right	Centre-Left/ Centre-Right
United Kingdom	Pro	Centre-Left	Centre-Left

Source: Column two was compiled out of the results from table 1 and the GAIN-Report 2009, 2009. Columns three and four were compiled from: (Bundeswahlleiter, 2010; BundesministeriumfürInneres, 2010; ComissãoNacional de Eleições, 2010; Congreso de los Diputados, 2010; Czech Statistical Office, 2010; Folkedinget, 2010; Hellenic Ministry of Interior, 2010; NLVerkiezingen, 2010; Panstwowa Komisja Wyborcza, 2010; Romanian Permanent Electoral Authority, 2010; Statistics Finland, 2010; The Electoral Commission, 2010; Valmyndigheten, 2010

on GMOs at their national levels, both concerning trade and cultivation (BMELV 2009).

This adds another dimension to the political and academic debate because so far it has not been acknowledged that countries who did not use the voting procedures at the EU-level to express their being against approval, also issued GMO bans at their own national levels. This further complicates multi-level governance. It is, however, beyond the scope of this chapter to further investigate the multi-level governance implications of this observation. In this cluster of Member States who consistently voted against the market authorisation of GMOs, we witness countries with strong green political movements and or parties, such as Austria, or green parties in government, such as Ireland and Germany. In the public debate, it is argued that 'Greens' always tend to be anti-GMO (The Greens – EFA 2011).

However, our analysis suggests that it is certainly not the rule that those governments including a green party vote against GMOs in EU level voting procedures. The German case demonstrates that despite the German Greens being part of the governing coalition in 2003–5, in none of the 18 votes did Germany explicitly oppose the introduction of a GMO. Portugal, Poland and Ireland have changed their view on the subject matter from pro or mixed attitudes towards anti-GMO stances after a change in government. However, this cannot be explained in terms of political ideology as the countries moved into different political directions. Ireland changed from a centre-right to a centre-green government, while in Poland and Portugal the centre-left was replaced by respectively centre-right and a social-democrat/right-wing coalition. Our analysis thus suggests that the voting behaviour of countries that are voting more or less consistently against the authorisation of GMOs and initiating bans at the Member State level cannot be understood in terms of political ideology, at least not in terms of the classic right-centre-left typology.

A Science-based Pro-GMO Stance?

According to the voting behaviour in EU procedures, there is also a cluster of Member States that opt for a more lenient and positive approach towards the introduction of GMOs on the EU market. Particularly the Netherlands, the Czech Republic, Estonia, Finland, (Ireland until 2007), Spain, Sweden and the United Kingdom belong to this group. The Netherlands, Finland, Sweden and the UK have voted positively on GMO issues in all or nearly all of the 18 analysed cases. Recently, the Dutch government, joined by the United Kingdom, even explicitly criticised the EU's system of market authorisation for GMOs. They claim that the system is too inefficient and rather propose a system whereby respective Member States would have the final word on whether a GMO may be grown on their soil (Reuters 2010). If the proposal is accepted, they hope to accelerate the procedure for their own countries and at the same time limit the degree of opposition faced on a EU-wide scale.

In the 2003 to 2005 period and afterwards also, the Netherlands, Finland and Sweden underwent changes in government, also in terms of political ideology, without any effect on their attitude towards GMOs. Furthermore, the political affiliations of the governments range from centre-left to centre-right, from social-democrat to liberal and green. So there is also not a consistent pattern that enables any explanation of pro-GM voting behaviour and political stances in classic political-ideology terms.

Mixed or Indecisive?

In their voting behaviour, a number of countries were ambiguous or ambivalent: France,[7] Belgium (not analysed in Table 6.2), Germany, Latvia (not analysed in Table 6.2), Italy, Poland (2003 to 2005) and to a lesser extent Portugal (until 2005). Can the inconsistency in voting behaviour be explained in terms of changes of government? Did these countries experience larger swings in the political affiliation of their governments? First of all, this would require a more detailed and precise analysis of the timing of the cases and the changes in government than we have carried out in the context of this chapter. However, at first glance it appears that the countries did not witness more radical changes in government than countries with a consistent voting pattern. Subsequent analyses should follow Jasanoff's (2007) approach in assessing a correlation between public opinion and political action but on a country-by-country analysis. In Italy for instance, public opinion is clearly against GMOs, which brings the government to be caught in the middle of opponents and proponents (industry). This seems to be the case in other Member States too (GAIN Report 2009; GMO-Free France 2010) but it deserves more analysis of the societal and political dynamics at the country level in order to be able to explain mixed voting behaviour as the upshot of being caught in the middle.

Political Ideology as an Explaining Factor?

To conclude, our data suggests that some countries have a fixed view on GMOs and an associated voting mentality and will not deviate from it, irrespective of changes in government. This holds for both proponents and opponents. However, neither the attitude in favour nor against GMOs can be explained in terms of political ideology or stability in the political affiliations of the government. For some countries (Poland, Portugal, Ireland) changes in voting behaviour seem to correlate with changes in government. However, the changes towards a more negative view on GMOs cannot be consistently explained in terms of the change in political ideology as the direction of the government change differs.

Conclusion

GMOs have proven to be a controversial topic in European societal and political debates. Analyses of the EU decision-making processes so far have concentrated on the bureaucratic level and the role of experts. In the academic literature, almost no attention has been paid to the political arena. Voting procedures are key in democratic societies and are thus critical for political analysis as they require taking a political stance. At the European Union level, such voting procedures on GMOs take place at the level of the standing committee, the SCFCAH, and should no qualified majority be reached, also on the Council level. In this chapter, we have examined the voting behaviour of the Member States in SCFAH and the Council in 18 GM authorisation decisions on 11 GMOs from 2003 to 2005. We compared our analysis with the GAIN Report from 2009. Quite the contrary to the expected transparency, it turned very difficult to map out the votes reliably. Underlying documents, such as minutes, turned out not to be publicly accessible. In all cases, the issue of trade was at stake as all of the GMOs involved included biotechnological agricultural products, such as herbicide-tolerant or insect-resistant maize and rapeseed crops produced outside the European Union. So the issue at stake was authorisation for trade with and within the European market.

The lack of qualified majority in all analysed authorisation decisions indicates maximum political controversy and broad diversity in the positions towards GMOs across the European Union. This implies that the European Union is actually divided into large minority voting blocks pro- and contra-GMOs. We also identified a third group expressing mixed voting behaviour. In the 2003 to 2005 period, some countries also abstained in nearly all cases. Due to tech-nocratic provisions, the ultimate decision to authorise the 11 GMOs in ques-tion was taken by the Commission. This not only harms the EU's ability to deal with biotechnologies but also raises issues of political legitimacy. The deadlock has led several Member States to invoke EU safeguard clauses so far, ban-ning GM crops that are officially acknowledged and approved for circulation throughout the whole of the European Union's territory. The European Union is often portrayed as being united in opposition to GMOs but our analysis indi-cates that it is actually due to the political inability to arrive at a qualified majority pro or contra, which has in fact provided the leeway for a technocratic approval of market authorisation by the European Commission. The intense political disagreement was not solved in the political arena. Against this background, it should come as no surprise that the politics moves to another level – in this case bans on the Member State level – which in turn give rise to further questions about multi-level governance in the European Union.

We attempted to account for the EU Member States' voting behaviour and the GMO positions revealed by their votes through analysing govern-ments in power, both in terms of changing electoral terms and political affiliations. We conclude that most Member States remain consistent in their stance towards GMOs, no matter whether the government has changed or not. We observed some changes in voting mentality towards a negative stance but none of the Member States moved in the other direction in the time period covered in our analysis. The landscape of EU national govern-ments in power is highly diversified with countries having leftist to far-rightist coalitions. It is not unusual that both have the very same opinion towards GMOs. Green parties, for example, do not automatically vote against the introduction of new GMO crops on the European level if they are in power. Thus, a consistent EU-wide pattern in terms of GMO position and political affiliation is not visible because a negative view on GMOs can be witnessed in centre-right, centre-left, social-democrat, liberal, conservative, green, far-right and centrist parties. Also the positive and mixed views cannot be understood in terms of a strong correlation with a particular political affiliation. Hence, governments' political affiliations do not explain their stance towards GMOs. This leads to the conclusion that political ideology in the classical right–left terminology is not useful to understand the positions on biotechnology of EU Member States and the question how to explain voting behaviour and revealed positions towards GMOs in political terms remains an open question. Based on our analysis, we rule out the explanatory value of the classic right–left political ideology scheme. It might be that more recent approaches for political analysis, such as Jasanoff's (2007) approach of political cultures and nation-building projects, may provide a better basis for understanding the politics of risk and trade.

The research described in this chapter is one of the first efforts to delve into and understand the political sphere and dimensions on EU decision making on genetically modified organisms. As the decision is oriented towards the market authorisation of products usually produced elsewhere, it can be considered an iconic example of risk-trade decision-making. Due to data availability and quality issues, only a fifth of those crops already authorised could be analysed, which implies that we should be careful with generalisations. Nevertheless, in our view, we have demonstrated the need to involve political analysis in the interdisciplinary research on risk and trade.

Notes

1 As shown by the case of the genetically modified maize MON810 this ideal situation is not always prevalent. With Germany banning MON810 from its market in 2009, six European Member States have now issued a ban against that GMO and thereby breaching the applicability assumption. This not only stands in opposition to the authorisation on behalf of the EU, but also triggered – based on the individual cases of Austria (1999) and Italy (2000) – a dispute in front of the WTO by the USA (GMO Compass 2011). See also Ivanova and Van Asselt in this volume.

2 Parts which could harm the applicant's economic interest and give away important innovative information to rivals will not be published.

3 Qualified majority voting (QMV) as it was outlined in the Treaty of Nice will apply until 31 October 2014 under the new Lisbon treaty. Each Member State is given a fixed number of votes, proportionate to its population (e.g. France 29, Spain 27). To pass a decision by QMV the following conditions must be satisfied: at least 14 (18 if proposal not made by the Commission) countries; at least 255 out of the total 345 voting weights; and – only if verification is requested by a Member State – at least 62 per cent of the total population of the EU (Art. 3, Treaty of Nice).

4 The new comitology rules will change this procedure somewhat in the near future. See the chapter by Kim *et al.* in this volume for more information on this.

5 The GAIN Report and Willis (2010) explicitly list Spain as a GMO-friendly country, although according to our data Spain has overwhelmingly abstained from votes and had mixed votes in the cases in which they voted.

6 Austria, Greece, Hungary and Luxembourg, as well as France and Germany.

7 Although according to the GAIN Report 2009, involvement of French regions in the alliance of GMO-free regions and bans on the Member State level indicate that France's position has become more critical in recent years.

References

Agrarheute (2011) 'Gentechnikgesetz: Verbände und Parteienmahnen Novellierung an', available online: <http://www.agrarheute.com/?redid=380008> (accessed on 31 March 2011).

Biosafety Information Centre (2004) 'Information on voting behaviour of Member States on GMO-related issues', available online: <http://www.biosafety-info.net/bioart.php?bid=200> (accessed on 22 May 2010).

BMELV (2009) 'Aigner: Entscheidung des Oberverwaltungsgerichts Lüneburg über Beschwerde von Monsanto', from Bundesministerium für Ernährung, Landwirtschaft und Verbraucherschutz, available online: <http://www.bmelv.de/cln_163/Shared Docs/Pressemitteilungen/2009/102-AI-MON810-Gerichtsbescheid.html> (accessed on 29 May 2010).

Borrás, S. (2006) 'Legitimate Governance of Risk at EU Level? The Case of GMOs', *Technological Forecasting and Social Change*, 73(1): 61–75.

Bundesministerium für Inneres (2010) 'Nationalratswahlen', available online: <http://www.bmi.gv.at/cms/BMI_wahlen/nationalrat/start.aspx> (accessed on 27 May 2010).

Bundeswahlleiter (2010) 'Ergebnisse der Wahlen auf Bundesebene', available online: <http://www.bundeswahlleiter.de/de/europawahlen/EU_BUND_09/ergebnisse/> (accessed on 28 May 2010).

Co-Extra (2010) 'GMO Compass Launches Online Discourse', available online: <http://www.coextra.eu/news/news396.html> (accessed on 28 May 2010).

Comissão Nacional de Eleições (2010) 'Eleiçãolegislativa', available online: <http://www.cne.pt/index.cfm?sec=0301000000> (accessed on 28 May 2010).

Congreso de los Diputados (2010) 'Resultado Electoral', available online: <http://www.congreso.es/portal/page/portal/Congreso/Congreso> (accessed on 29 May 2010).

Czech Statistical Office (2010) 'Results of Elections', from Czech Statistical Office, available online: <http://www.volby.cz/index_en.htm > (accessed on 29 May 2010).

EurLex (2004) 'Commission authorises import of canned GM-sweet corn under new strict labelling conditions – consumers can choose', available online: <http://eur-lex.europa.eu/LexUriServ/LexUriServ.do?uri=CELEX:52004PC0439:EN:HTML> (accessed on 24 May 2010).

Eurobarometer (2006) 'Europeans and Biotechnology in 2005: Patterns and Trends', available online: <http://ec.europa.eu/research/press/2006/pdf/pr1906_eb_64_3_final_report-may2006_en.pdf > (accessed on 4 April 2011).

European Commission (2010) 'GMOs: Commissioner Dalli receives a petition from Greenpeace and Avaaz', available online: <http://ec.europa.eu/commission_2010–14/dalli/docs/GMO_Midday_GreenpeacePetition_09122010_en.pdf > (accessed on 9 April 2011).

Europa (19 May 2004) 'Commission Authorises Import of Canned GM-sweet Corn under New Strict Labelling Conditions – Consumers Can Choose', available online: <http://europa.eu/rapid/pressReleasesAction.do?reference=IP/04/663&format=HTML &aged=0& language=EN&guiLanguage=en> (accessed on 18 May 2010).

FOE-Europe (2010) 'Table on how the EU Member States Voted on GMOs?', available online: <http://www.foeeurope.org/GMOs/pending/votes_results.htm> (accessed on 22 May 2010).

——(2011) 'GMOs, Food and Farming Campaign. Working for Healthier, Tastier Food that is Good for the Environment and People', available online: <http://www.foeeurope.org/GMOs/Index.htm> (accessed on 3 April 2011).

FOE-UK (2004) 'Monsanto GM Maize Not Authorised by EU Commission', available online: <http://www.foe.co.uk/resource/press_releases/monsanto_gm_maize_not_auth_28062004.html> (accessed on 24 May 2010).

Folkedinget (2010) 'Danish Election Results', available online: <http://www.ft.dk/> (accessed on 26 May 2010).

Follesdal, A. and Hix, S. (2005) 'Why there is a Democratic Deficit in the EU', European Governance Papers (EUROGOV) No. C-05-02, available online: <http://www.connex-network.org/eurogov/pdf/egp-connex-C-05-02.pdf> (accessed on 1 October 2011).

GAIN Report 2009 (2009) 'EU-27 Agricultural Biotechnology Annual 2009', available online: <http://www.seedquest.com/News/pdf/2009/GAIN_EU_Biotech.pdf> (accessed in May 2010).

Gaskell, G., Allum, N., Wagner, W., Gaskell, G., Kronberger, N., Torgersen, H., Hampel, J. and Bardes, J. (2004) 'GM Foods and the Misperception of Risk Perception', *Risk Analysis*, 24: 185–94.

GMO Compass (2011), 'Food and Feed from GMOs – The Long Road to Authorisation', available online: <http://www.GMO Compass.org/eng/regulation/regulatory_process/157.eu_gmo_authorisation_procedures.html> (accessed on 19 May 2011).

GMO-free Europe (2010) 'Maps of GMO-free Zones in Europe', available online: <http://www.gmo-free-regions.org/gmo-free-regions/maps.html> (accessed 24 May 2010).

GMO-free France (2010) 'GMO-free Zones in France', available online: <http://www.gmo-free-regions.org/gmo-free-regions/france.html> (accessed 24 May 2010).

Greenpeace (2011a) 'Gentechnik', available online: <http://www.greenpeace.de/themen/gentechnik/> (accessed 29 March 2011).

——(2011b) 'Mach_dich_vom_Acker', available online: <https://service.greenpeace.de/themen/gentechnik/kampagnen/mach_dich_vom_acker/> (accessed 29 March 2011).

Hellenic Ministry of Interior (2010) 'Election Results', available online: <http://www.ypes.gr/en/Elections/NationalElections/Results/> (accessed 28 May 2010).

Hunter, R. (1999) 'European Regulation of Genetically Modified Organisms', in J. Morris and R. Bate (eds), *Fearing Food: Risk, Health and Environment*, Newton, MA: Butterworth-Heinemann, pp. 189–230.

ISIS (2004) 'Europe Still Resisting GMOs', available online from the Institute of Science in Society: <http://www.i-sis.org.uk/ESRG.php> (accessed 26 May 2010).

Jasanoff, S. (2007) *Designs on Nature: Science and Democracy in Europe and the United States*, Princeton: Princeton University Press.

Levidov, L. (2001) 'Genetically Modified Crops: What Transboundary Harmonization in Europe?', in J. Linnerooth-Bayer, R.E. Löfstedt and G. Sjöstedt (eds), *Transboundary Risk Management*, London: Earthscan Publications, 59–90.

Löfstedt, R.E., Fischhoff, B. and Fischhoff, I.R. (2004) 'Precautionary Principles: General Definitions and Specific Applications to Genetically Modified Organisms', *Journal of Policy Analysis and Management*, 21(3): 381–407.

Majone, G.D. (2005) *Dilemmas of European Integration. The Ambiguities and Pitfalls of Integration by Stealth*, Oxford: Oxford University Press.

Marchant, R. (2001) 'From the Test Tube to the Table', *EMBO Rep.*, 2: 354–57.

NLVerkiezingen (2010) 'Nederlandse verkiezingsuitslagen 1918-nu', available online: <http://www.nlverkiezingen.com/> (accessed on 27 May 2010).

Nowotny, H., Scott, P. and Gibbons, M. (2001) *Re-thinking Science: Knowledge and the Public in an Age of Uncertainty*, Cambridge, UK: Polity Press in association with Blackwell Publishers.

Panstwowa Komisja Wyborcza (2010) 'Polish Election Results Sejm', available online: <http://www.pkw.gov.pl/pkw2/index.jsp?place=Menu01&news_cat_id=30&layout=0> (accessed on 29 May 2010).

Pidgeon, N., Kasperson, R.E. and Slovic, P. (2003) *The Social Amplification of Risk*, Cambridge: Cambridge University Press.

Reuters (2010) 'FACTBOX-How EU Member States Approach GMOs', available online: <http://www.reuters.com/article/idUSLDE6120K120100204> (accessed on 24 May 2010).

Romanian Permanent Electoral Authority (2010) 'Election Statistics', available online: <http://www.roaep.ro/en/section.php?id=66> (accessed on 28 May 2010).

SOS (2004) 'Votes at the Regulatory Committee of 2001/18/EC 29 November 2004', available online save our seeds: <http://www.saveourseeds.org/fileadmin/files/SOS/Dossiers/Mon810/results_nov_29-11-04.pdf> (accessed on 27 May 2010).

Statistics Finland (2010) 'Statistics on Different Elections', available online: <http://www.stat.fi/tk/he/vaalit/index_en.html> (accessed on 25 May 2010).

The Electoral Commission (2010) 'Results and analysis', available online: <http://www.electoralcommission.org.uk/elections/results> (accessed on 29 May 2010).

Tulloch, J. and Lupton, D. (2002) 'Consuming Risk, Consuming Science: The Case of GM Foods', *Journal of Consumer Culture*, 2(3): 363–83.

The Greens – EFA (2011) 'Who we are – History', available online: <http://archive.greens-efa.eu/cms/default/rubrik/6/6648.history@en.htm> (accessed on 10 July 2011).

Valmyndigheten (2010) 'Swedish Election Results', available online: <http://www.val.se/tidigare_val/index.html> (accessed on 29 May 2010).

Van Asselt, M.B.A. and Vos, E. (2006) 'The Precautionary Principle and the Uncertainty Paradox', *Journal of Risk Research*, 9(4): 313–36.

——(2008) 'Wrestling with Uncertainty Risks: EU Regulation of GMOs and the Uncertainty Paradox', *Journal of Risk Research*, 11(1–2): 281–300.

Vos, E. and Wendler, F. (2006) 'Food Safety Regulation at the EU Level', in E. Vos and F. Wendler (eds), *Food Safety Regulation in Europe. A Comparative Institutional Analysis*, Mortsel, Belgium: Intersentia Publishing, pp. 65–138.

Willis, A. (2010) 'Spain a Key Ally of pro GMO-America, Cables Reveal', *EU Observer*, available online: <http://euobserver.com/891/31544> (accessed on 4 May 2011).

7 Regulating the Use of Bisphenol A in Baby and Children's Products in the European Union

Trade Implications of an Uncertain Risk

Tessa Fox, Esther Versluis and Marjolein B.A. van Asselt

Introduction

Parents of infants have recently been confronted with labels indicating that their purchases of baby bottles, teethers or sippy cups are now 'Bisphenol A-free'. What does this imply? These labels suggest that the presence of Bisphenol A (BPA) poses a risk of which consumers, most likely, were unaware. Should parents seek information on this BPA-free label? The product instructions do not elaborate on its meaning or rationale. Searching online, however, would expose a wealth of information about the possible risks and uncertainties surrounding products containing BPA. It is a sensitive topic of debate as such products are intended for one of the most vulnerable groups within society: children under the age of three.[1]

BPA is systematically framed as a risk by the media. It became arguably the new buzzword in media coverage on chemical risk regulation (Ward Barber 2010; Lozito 2007). Bisphenol A, a synthetic chemical used in the production process of polycarbonate (plastics) worldwide, is making headline news in the US and the EU. Its questioned safety in materials in contact with foodstuffs, baby bottles and children's toys has made plastics a political issue (Vogel 2009).

In the past few years, experts, NGOs, industry, retailers and regulators have all been wrestling with the question of how to deal with BPA and whether it may be used in children's and baby products. At the heart of this debate lies scientific uncertainty regarding the possible risks of the substance, through leakage or dosage. NGOs and several scientists have been requesting a review of existing exposure thresholds set within the EU and the US. Recently various risk assessment reports issued both in the US (EPA 2007; FDA 2007; National Toxicology Programme 2007) and several EU Member States (VWA 2006; Danish Ministry of Food and Agriculture 2010; FSA (UK) 2010) as well as the EU (EFSA 2006), have argued in favour of existing regulatory limits to BPA exposure. The EU's regulatory framework for chemicals, REACH,[2] requires the registration of all existing and new chemicals, which have to be subjected to a risk assessment review. As such,

Media coverage addressing BPA controversy

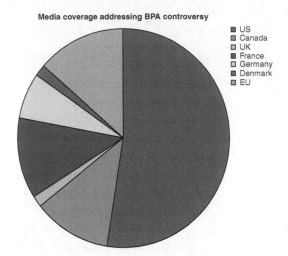

Figure 7.1 Illustration international media coverage on BPA between 2008–10

BPA as a chemical will fall under the requirements set by REACH. In addition, the continuous media coverage illustrating scientific uncertainty around the possible adverse health effects[3] seems to have put pressure on the actors involved to request regulatory action (Kitzinger 1999; Majone 2010; Pidgeon, Kasperson and Slovic 2003). The challenge therefore is to regulate the use of a chemical like BPA, particularly in the light of scientific uncertainty about its possible risks and the various trade implications that may arise from regulatory action.

This contribution aims to analyse whether and how the EU should regulate the use of BPA in baby products, taking into account the lack of certainty and possible trade implications. Using documentation sources and media analyses and interviews,[4] the main purpose here is to gain an understanding of how the European Union approaches regulation in the face of uncertainty surrounding risks. To this end, the self-regulatory behaviour of industry is also considered. On the basis of the current regulatory tools in place and the current state of affairs, as well as developments in the controversy around BPA, this contribution will present four scenarios, each depicting a possible regulatory future for Bisphenol A in the EU.

Setting the Scene: Bisphenol A and Scientific Uncertainty

Many toys and bathing products for young children contain plastics. In these plastics, chemicals such as phthalates and BPA are used alongside many others in order to optimise the manufacturing process. It is also true that young children tend to put toys into their mouths and nibble on them. It is feared that these chemicals may end up in the body of the infant or toddler and that this may cause neurological and physical damage such as learning disabilities and hormonal interferences. Risk assessment agencies have carried out risk

Table 7.1 Selection of articles in serious media in various countries

Headline	Year	Newspaper/ magazine	Source
'The Perils of Plastic'	2010	Time Magazine (US)	Bryan Walsh (2010) The Perils of Plastic. Thursday, Apr. 01, 2010 http://www.time.com
'A truly toxic issue'	2010	The Guardian (UK)	Sadhbh Walshe (2010) A truly toxic issue. guardian.co.uk, Saturday July 10 2010 http://www.guardian.co.uk/commentisfree/cif-green/2010/jul/10/avoid-hazardous-chemicals
'Bisphenol A banned in baby bottles'	2010	Le Monde (FR)	AFP (2010) Le bisphénol A définitivement banni des biberons. LEMONDE.FR10.05.10 on: http://www.lemonde.fr
'Bisphenol A : Experts acknowledge the existence of warning signals'	2010	Le Monde (FR)	Dupont, G. (2010). Bisphénol A : des experts reconnaissent l'existence de signaux d'alerte 05 February 2010. Source : LE MONDE.FR
'Toxic baby bottles'	2010	Le Figaro (FR)	Cusso, M. (2010). Les biberons toxiques23 janvier 2010 http://recherche.lefigaro.fr
'Back to using glass baby bottles'	2010	Kids Feeling (NL)	Kids Feeling, Terug naar glazen babyflessen May 2010, http://www.kidsfeeling.nl
1. 'Revealed: the nasty secret in your kitchen cupboard'	2010	The Independent (UK)	1. Hickman, M. (2010). Revelealed: the nasty secret in your kitchen cupboard. April 2010 on http://www.independent.co.uk
2. 'Boots vows to phase out baby bottles containing chemical'	2010	The Independent (UK)	2. Hickman, M. (2010). Boots vows to phase out baby bottles containing chemical. April 2010 on http://www.independent.co.uk
3. 'Scandal of danger chemical in baby bottles'	2010	The Independent (UK)	3. Hickman, M. (2010). Scandal of danger chemical in baby bottles. March 2010 on http://www.independent.co.uk

Table 7.1 (continued)

Headline	Year	Newspaper/ magazine	Source
'Canada Becomes First Country to Officially Declare BPA a Health Hazard'	2009	Treehugger (online blog)	Merchant, B. (2009) Canada Becomes First Country to Officially Declare BPA a Health Hazard 04.14.09 on: http://www.treehugger.com/files/2009/04/canada-bpa-health-hazard.php.
'Bisphenol A in pacifiers: sucking with side effects'	2009	Die Zeit (Germany)	Ludemann, D. (2009). Bisphenol A in Schnudllern – Nuckeln mit Nebenwirkungen 15.10.2009: http://www.zeit.de/wissen/2009-10/bisphenol-schnuller-gesundheit
'Straight through the EU rules and regulations'	2009	Arbejderen (DK)	Sondergaard, S. (2009). Tvars igennem EU's love og regler. On: http://www.arbejderen.dk/artikel/2009-12-15/tv-rs-igennem-eus-love-og-regler
1. 'Reassessing the Dangers of BPA in Plastics'	2008	Time Magazine (US)	1. Park, A. (2008) Reassessing the Dangers of BPA in Plastics Sunday, Nov. 02, 2008 http://www.time.com
1. 'Bisphenol A is linked to heart disease'	2008	The Guardian (UK)	1. (2008). Bisphenol A is linked to heart disease. September 16, 2008 on:http://www.guardian.co.uk
2. 'Risky Business'	2008	The Guardian (UK)	2. Miller, H. (2008). Risky Business. July 11, 2008 on: http://www.guardian.co.uk
3. 'Hot water increases baby bottle chemicals'	2008	The Guardian (UK)	3. Jha, A. (2008) Hot water increases in baby bottle chemicals. January 30, 2008 on: http://www.guardian.co.uk

Source: T. Fox

assessments which conclude that there are no indications of harm or adverse health effects when BPA containing products are used as intended (EPA 2007; VWA 2006; Danish Ministry of Food and Agriculture 2010). On the other hand, in the period 2007–9, studies on BPA and Phthalates (Health Canada 2008; Vom Saal and Myers 2008) have highlighted adverse health effects. Such studies increased controversy that has already surrounded the topic for decades. Under EU law,[5] certain categories of phthalates have been prohibited in the manufacturing of toys used by children under the age of three. Therefore, questions remain as to the use of BPA in toys and other products used by children. BPA ranks in the top 3 per cent of high production volume chemicals and is a synthetic compound commonly used in the production of polycarbonate. It is a plastic applied in the automotive industry, in the production of CDs, bicycle helmets, electronic devices, food cans, wrapping materials for soft drinks, and children's products, such as sippy cups and toys. Baby bottles and children's products fall within bottles and food packaging, which collectively represent only 3 per cent of the annual production volume applications of BPA. Sippy cups are usually made out of polycarbonate because they can be cleaned easily and are almost unbreakable (Babyplaza 2010).

Since the 1930s, BPA has been a rather controversial chemical. After its invention in 1891, its estrogenic properties were discovered in the 1930s by Dodds and Lawson (1938). These findings were considered valuable for decreasing the risks of miscarriage. Although considered for pharmaceutical use in the early 1930s, BPA has never been used due to the more effective estrogenic properties of another synthetic chemical, diethylstilbestrol (DES). This compound was taken off the market in the 1970s after birth defects and other adverse health effects for newborn girls were linked to the use of DES during pregnancy (Harremoës, Gee, MacGarvin et al. 2002; Meyers 1983; Apfel and Fisher 1984; Ferguson 2002). This negative association with DES also affects the risk perception concerning BPA.

From the 1940s onwards, BPA has been used in the manufacture of plastics in ever-increasing volumes. In the US alone, production volumes approach 45 billion kilograms annually. Despite occasional protests by NGOs, it was not until the 1990s that actual adverse effects related to dosage were found in lab studies (Erickson 2008). Since then, various scientific reports have found that chronic use of BPA might cause severe health problems. Although BPA is said to have low acute toxicity, it might also be a so-called 'endocrine disruptor', which, according to some scientists, could cause adverse health effects in cases of chronic exposure. The endocrine system is one of the two bodily communication systems and it 'transports information through the circulation from one tissue to another via hormones' (Witorsch 2002). Endocrine disruptors are human-made synthetic chemicals that 'mimic, block and/or interfere in some manner with the natural instructions of hormones and cells' (Goettlich n.d.: 8). The concept was coined in 1991 at the Wingspread Conference Centre in Wisconsin (Vogel 2009) where a diverse group of leading scientists (endocrinologists, toxicologists, biologists and physiologists among others) concluded that certain chemicals are able to interfere with human endocrine systems and the environment. This gathering was organised after a 1993 Stanford research endeavour by endocrinologists had identified a BPA leakage from polycarbonate in their laboratory (Vogel 2009).

Since 1993, the field has been in development. Frederick vom Saal, curators' Professor at the University of Missouri-Columbia (US), has published numerous articles in highly rated journals and is considered to be a leading scientist in the field of biological sciences. He focuses on 'low dose, in-vivo effects of molecular mechanisms based primarily on in vitro studies, sources of exposure and pharmacokinetics relating to Bisphenol A' (Vom Saal and Myers 2008). As part of the Endocrine Disruptors group that was founded at the University of Missouri, Vom Saal takes the lead among a group of scientists who argue that long-term consequences of chronic exposure include possible behavioural problems, learning disabilities, brain defects, early age puberty, carcinogenic effects and other hormonal issues (Murray, Maffini, Ucci et al. 2007). However, scientific conclusiveness on this issue is still lacking.

Scientific Inconclusiveness

There has been a great deal of debate among regulators, NGOs, consumer groups and experts on the uncertainty of adverse health effects of BPA. Despite a gradual increase in scientific research concluding that BPA poses a risk, the amount of literature contesting this finding seems equally abundant and of similar stature. Scientific inconclusiveness has dominated debates lasting decades on the possible risks of BPA. These debates have centred on: i) dosage; ii) possible migration (or leakage) of the substance, and iii) its interaction with other chemicals.

With regard to dosage, various scientific reports (Carr, Bertasi, Betancourt et al. 2003; Honma, Suzuki, Buchanan et al. 2002; Negishi, Kawasaki, Suzaki et al. 2004; Vom Saal, Nagel, Timms et al. 2005; Vom Saal and Hughes 2005) claim that dosage is no longer immanent to the possible danger a substance may pose. It is argued that infants and toddlers chronically exposed to small doses of BPA may also be subject to long-term adverse health effects, which include learning disabilities and hormonal disturbances. These scientists refer to lab studies showing that low dosage exposure to BPA had adverse effect on mice and rats.[6]

On the other hand, other experts (Gray, Cohen, Cunha et al. 2004) emphasise that 'dosage makes the poison' and this would suggest that only acute exposure to high levels of BPA would jeopardise children's health rather than the current application levels of the substance in products which are below certain regulatory exposure thresholds. These experts argue that lab studies suggesting adverse effects at low dosage exposure are not fully reliable since animals can respond differently from human beings, especially to substances which may have hormonal effects.[7]

Uncertainties pertaining to the possible migration of BPA is of particular concern with regard to BPA in toys. The issue of migration plays out on two levels: firstly, whether any release or migration of BPA may occur when heating a baby bottle or scraping its surface, such as when a baby nibbles on a bottle, teether or pacifier (Lamb 2002; Kubwabo, Kosarac, Stewart *et al.* 2009; Vom Saal, Nagel, Timms *et al.* 2005); secondly, whether or not this migration process poses health risks to the infants and small children (Vandenberg, Hauser, Marcus *et al.* 2007; Mountfort, Kelly, Jickells *et al.* 1997).

Thirdly, interactions of BPA with other chemicals are also contentious. Since there are hundreds of thousands of chemicals that have not been assessed, some regulators[8] and scientists argue that it is impossible that a risk of this nature can ever be ruled out. Hence they argue that this interaction risk is invalid and baseless:

> When you have to start looking for combinations that may cause harm due to the interaction with BPA, then there are infinite possibilities, as there would be with any chemical, and hence ruling that risk out is simply impossible.[9]

However, various NGOs, regulators and scientists remain convinced of the need to consider this risk. Some scientists argue that BPA is an endocrine disruptor, and that general risk assessments are inadequate since they overlook the possibility that BPA may bind oestrogen receptors and may cause transactivation of oestrogen-responsive genes. Additionally, they recommend a different type of risk assessment in order to account for these interactions (Safe, Pallaroni, Yoon *et al.* 2002).

Furthermore, it seems that the largest uncertainty regarding the possible effects of BPA resides in the interpretation and reliability of scientific research. Scientific uncertainty seems to be perpetuated by the multi-actor situation pertaining to BPA. Industry (both the chemical BPA manufacturer and the childrens' toys industry), scientists, regulators, NGOs and consumers all contribute different expertise, objectives, questions and 'evidence' to the debate. The reliability of this 'evidence' is a sensitive component in the risk dossier and deserves additional consideration; as is illustrated in the next part of this contribution.

In 2009, Sarah Vogel reviewed publications (mainly US- and EU-based) on the 'safety' of BPA. Her research indicates that the scientific attention to BPA, especially in the field of low-dose effects of endocrine disruptors, has expanded during the past two decades. According to Vogel, 115 studies have been conducted in Japan, the US and the EU between 1997 and 2005, looking into low-dose effects. Vogel argues that a vast majority of 104 studies indicate inconsistencies which collectively challenge the presumption that BPA has no low-dose hormonal effects (Vogel 2009: Honma, Suzuki, Buchanan *et al.* 2002; Negishi, Kawasaki, Suzaki *et al.* 2004). Other studies – 11 according to Vogel, among which is a large multigenerational study conducted by the Harvard Centre for Risk Analysis – have found no inconsistencies and have 'cast doubt on suggestions of significant physical or functional impairment' (Gray, Cohen, Cunha *et al.* 2004).

BPA as an Uncertain Risk

The scientific uncertainty about the effects of BPA on human health in general, and the possible larger effects on children in particular, is thus obvious. Even in attempts to suggest scientific conclusiveness with regard to the risk (or actual safety, hence the non-existence of significant risks), scientific uncertainty is visible through disagreement within the group of experts responsible for the risk assessment. Due to this state of uncertainty, we argue that BPA should be qualified as an uncertain risk. The notion of uncertain risks is used, for example, by Van Asselt and Vos (2006) to refer to risks that may be:

> distinguished from 'safe uncertainties', as uncertain risks pertain to uncertain situations, which may result in one or more effects, which are valued negatively or considered unacceptable by at least one, but possibly more, societal actors
>
> Van Asselt and Vos 2006

Renn (IRGC 2005) proposes to further differentiate between uncertain risk, complex risks and ambiguous risks. However, Versluis *et al.* (2011) argue that these are not independent categories and that the most important distinction is the difference between simple risks and uncertain risks (Van Asselt and Renn 2011; Hermans *et al.* 2012). At the core of this distinction lies the fundamental difference in assessment, management and communication tools needed to address the different types of risk (Wynne 2001; WRR 2008; Löfstedt 2005; Gezondheidsraad 2008; IRGC 2005).

In some cases, such as car accidents, risks are simple and probability and effect may be determined by using a 'positivistic' approach (Krayer von Kraus 2005; Van Asselt 2000). However, it has been argued that these tools no longer suffice when dealing with uncertain risks. The risks and their effects cannot be predetermined due to the inherent uncertainty. In recent years, it has been proposed to switch to a new way of dealing with risk, generally referred to as risk governance. By now, the contours of a framework of risk governance have been developed.[10] The classic positivistic risk approach is based on two phases – risk assessment and risk management – which together aim to assess the acceptability of risk (WRR 2008) by means of statistical calculations where uncertainty seems to play no role or it is difficult to adequately consider radical uncertainty. Risk governance, on the other hand, puts uncertainty centre stage and includes a multi-actor, multi-faceted risk process including contextual factors which together determine the roles, relationships and responsibilities of particular actors and mechanisms (Renn and Walker 2008). This means that qualifying a particular risk dossier in terms of uncertain risks has regulatory consequences.

BPA and the EU

The scientific controversy surrounding BPA is also reflected at the EU regulatory level. Since the 1990s both regulators and scientists have been divided over the possible risks of BPA. At the EU level, the EU Commission's Joint Research Centre conducted EU Risk Assessment Reports in 2003 and 2008 and reviewed hundreds of studies (European Chemicals Bureau 2003, 2008) of BPA with regard to migration, leakage or dosage, and which have highlighted 'some concern'. Obviously, this list was non-exhaustive. Subsequent interpretation by the EU's risk assessment committee in the JRC led to most of the research on BPA being dismissed.

Overall, the EU risk assessment committee found that prior research fell into three categories:

- Recent research efforts did not yield any new scientific evidence compared to prior assessments according to the panel. Hence, BPA does not constitute a risk according to the panel;
- A majority of the studies which were evaluated were considered to be incomparable for review by the panel and it concluded that: 'Consistency assessment[11] shows that there is no discernable and reproducible pattern to the behavioural testing results. Most of the studies reviewed by the panel have not been replicated at independent laboratories, so consistency cannot be assessed'[12]; and
- The concern of Norway, Denmark and Sweden about the disqualification of several studies by the panel was represented in a footnote, but seemed to have been dismissed when looking at the committee's overall conclusions.

The committee opted to disqualify most of the evaluated research:

> Taking together the low confidence in the reliability of developmental neurotoxicity studies and the lack of [replication availability], no conclusions can be drawn from these studies (this is in line with the European Food Safety Authority [EFSA] conclusions of 2006)

This suggests that the studies relied on for this risk assessment do not warrant a regulatory response or an alteration of current exposure limits.

However, there are some cracks in the collective position of the EU committee members. A footnote on page 120 indicates that experts from Norway, Sweden and Denmark did not agree with the panel conclusions. They find the methodology, group sizes and databases used by several researchers 'sufficiently reliable for regulatory use'. The three countries' experts voiced a preference for a twofold conclusion:

- That the available but limited data be used in the risk assessment, and
- That there is a need for further information after this risk assessment has been carried out.

This disagreement relates to the section on brain morphology. However, similar objections are raised throughout the document whenever the effects of developmental neurotoxicity are discussed. Norway, Sweden and Denmark state:

> The effects found in these studies[13] indicate that there is a possible risk for developmental neurotoxicity of BPA at very low exposure levels (0.1–0.25 mg/kg/day). These effects cannot be dismissed on the basis of other unreliable studies
>
> ECB 2008: 120

It seems, however, that both the suggestions for including these reports in the risk assessment as well as the call for further information were not taken up by the panel. The conclusions for both 'humans exposed via the environment' as well as all 'consumer exposure scenarios', read: 'There is at present no need for further information and or testing or for risk reduction measures beyond those which are being applied already' (EJRC 2010).

Scientific uncertainty thus remains dominant in the discussion of possible adverse health effects of BPA. In 2010, the EJRC published an overview of the risk assessments that have been carried out at the international level over the past 10 years. In its conclusions, the authors refer to a need for further research co-ordinated at the international level with special attention to the risks of substitutes given that industry is now voluntarily renouncing BPA products prior to regulatory action (EJRC 2010).

The EU Regulatory Framework

Although the EU has relatively recently engaged in the regulation of chemical substances it is allegedly setting an international example[14] with the implementation of the extensive procedural design of the REACH framework (Freshfields Bruckhaus Deringer 2010; Majone 2010; Bowman and Van Calster 2007). Prior to the 1960s, the admission of chemical substances into the European Union was practically unregulated. Since the 1960s, the EU has adopted regulations that cover the particular issues of labelling and the classification of chemical substances. This policy was in place for decades until a 1990 policy review indicated several policy impediments for the development of new chemicals. These new chemicals would only receive market entry permission after fulfilling testing requirements whereas existing chemicals, placed on a register in 1981, received market entry without having to fulfil requirements. This led to unfair competition and impeded innovation.

Furthermore, procedures were manifold and were executed at a slow pace, mainly due to the fact that national governments were the sole executors and were tasked with the risk assessment, management and communication requirements. The workload for the EU Commission and the individual Member States consisted of obtaining data on the usage, exposure and possible hazards of a particular chemical. On the basis of the collected information and evaluation they had to develop concrete management procedures. With the implementation of the REACH framework in 2007[15] a large part of this burden has now been shifted to industry, requiring companies to register the chemical substances they produce accompanied by a risk assessment. Hence, part of government's role as a risk assessor is shared with industry. Furthermore, risk assessment will only be conducted in cases of requests by a Member State or the European Commission.

One of the main goals of REACH is to provide sufficient information to be able to assess the potential risks of chemical substances. REACH requires manufacturers and importers of chemical substances to register their uses and categorises these substances according to the risk they pose to human health and the environment.

Under REACH, 'substances of very high concern' may be subject to 'authorisation' in order to ensure that the risks from these substances are properly controlled.[16] Based on the information from industry, the Member State committee within the European Chemicals Agency (ECHA) will evaluate which chemicals should be nominated to appear on a list subject to authorisation. Substances classified as CRM (Carcinogens, Reproductive Toxins and Mutagens) or PBTs (Persistent, Bio-accumulating and Toxic substances) will automatically be nominated for restriction. This list will be presented to the European Commission as an advice, whereupon the latter has the decision making powers to adopt the short list after having consulted a committee composed of national representatives typically known as 'comitology' committees.[17] The last stage is potentially crucial since it is open to lobbying by NGOs and industry and the influence of the European Parliament, although NGOs state that they do not have full access to comitology processes and procedures, and that comitology is still obscure.

Chemicals that require prior approval by the European Commission by means of an authorisation, face two possibilities: i) the adequate control route, which means that the chemical will be restricted in its use, or ii) the socio-economic route, which indicates that benefits outweigh the risks. The ECHA manages the technical, administrative and scientific aspects of REACH and the Commission can cancel restrictions whenever 'suitable substitutes' become available.[18]

Regulatory Mechanisms

The regulatory process of chemical applications is not straightforward. Chemicals are not only subject to EU regulation but are also used on the global markets and therefore also have a considerable potential for triggering international trade disputes. Hence, one should not only distinguish between the regulatory regimes of various countries (Majone 2010; Wiener, Rogers and Sand 2010) where the EU, US and Canada are relevant[19] but should also take account of the fact that actors within these different regulatory regimes tend to respond differently in various cases, since each response is determined by the particularities of the context in which the actor operates. For example, the regulatory process of a chemical such as BPA will be approached differently by industry than by members of the Risk Assessment Committee of the European Chemicals Agency (ECHA), since they each have different goals and accompanying job descriptions. Consequently, regulation of such substances, staged in a controversy that is framed in terms of uncertain health risks, is a complex issue.

BPA is not the first chemical substance that has stirred up debate on food and health safety for adults and children. Phthalates, esters of phthalic acid,[20] are the most commonly used plasticisers in the world. They are used for a wide array of products, ranging from medical care products and building materials to children's toys. The types of phthalates used in the production of children's products are DIDP, DNOP, DINP, BBP, DBP and DEHP. These substances also have been headline news in the US and EU prior to BPA and currently, depending on the geographic location concerned, either have been subjected to strict regulation (in the EU), are being phased out (in the US), or are threatened with a regulatory ban (in Canada).

What has been the motivation behind the restrictive measures in the case of the particular types of phthalates? Heudorf *et al.* (2007) and Sathyanarayana *et al.* (2008) suggest that infants experience a larger and more frequent exposure to phthalates than adults due to its presence in lotions, shampoos, teethers and toys. Since 1999 the EU has imposed restrictions on the use of some phthalates in the European Union for use in children's toys:

> The placing on the European market of childcare articles and toys in soft PVC containing phthalates which are intended to be put in the mouth by babies under 3 years, should be banned immediately. The European Commission has approved a package of legislative proposals to ban the use of phthalates in soft toys. Recent scientific evidence has shown that checks of phthalates release from such products in soft PVC are not a reliable way to ensure that babies do not absorb dangerous quantities of phthalates.[21]

Thus, DEHP, BBP, and DBP are restricted for use in all toys.[22] DINP, DIDP, and DNOP are restricted only in toys that can be taken into the mouth, such as teethers and yellow duckies. The restriction was recommended by the Scientific Committee on Food (SCF 2002) and stipulates that the percentage of phthalates may not represent 0.1 mass per cent of the plastic parts of the toy.

Phthalates are currently part of the pre-registration phase of the Candidate List of Dangerous Substances. REACH entered into force on 1 June 2007. Due to pending court cases, initiated by producers of the chemical acralite, which was also nominated, this process has experienced considerable delay. Another setback has been caused by 'freezing relations between the two different captains on one ship', referring to DG Enterprise and DG Environment within the EU Commission.[23] After all, a BPA ban may have serious trade implications.

After almost a year, the Commission has forwarded a priority list of six substances to the EP, for review until January 2011. Representatives of NGOs as well as some experts involved in substance evaluation[24] envisaged that the results of the process to which phthalates was subjected, would serve as the precedent for BPA.

However, the regulatory treatment of BPA does not resemble that of phthalates. BPA has allegedly been subject to a more precautionary approach in the US than in the EU. In the US, the potential risks of BPA have been intensively covered by the media, raising controversy nationwide. Both scientific uncertainty and the role of industry have added to the risk controversy. In this context, ten US states have banned the use of BPA, and more may be expected to follow. This does not imply a federal ban in the US, since there is not enough support for such an initiative. However, the latest US FDA risk assessment update has indicated 'some concerns' of possible adverse health effects of BPA, following a sequence of past risk assessments refuting any health concerns. In the EU, the Commission previously argued that risk assessments carried out by Member States found that BPA poses 'no risk' to human health or the environment. Current levels of BPA in the products concerned are allegedly within the set safety margin of 50 mg/kg/day.

Moreover, the EFSA did not amend its 'Tolerable exposure level' after the first round of assessments carried out by the EJRC. In July 2010, the results of a new risk assessment of BPA by EFSA, expected in the summer of 2010, were delayed until the September. An evaluation of more than 800 studies followed after the advice of the Danish DTU Food Institute on the risk assessment of BPA, led to a ban of its use in Denmark shortly afterwards in 2010. The EFSA panel has now recommended that in their deliberations to date 'they would like to maintain the Tolerable Daily Intake (TDI) for BPA at 0.05 mg/kg body weight per day', although some suggestion was made to convert the TDI to a temporary TDI.

Although there is 'no evidence' that would lead to a change of the TDI, at the same time, the Panel has identified areas of uncertainty which merit further consideration. Sources at the Dutch National Authority for Food Safety revealed that a January 2009 draft version of this report proposed no changes to current regulatory limits and requirements to the use of BPA. When the final version was still pending, it fuelled hope with NGOs and several scientists that a pending answer is an acknowledgement of 'sufficient' uncertainty about possible effects to support a phasing out of BPA in food and children's products. EFSA's risk assessment report could initially lead to a nomination for restriction by concluding, in line with the US Food and Drug Administration in 2009, that there is 'some concern' as to the use of BPA products in food packaging, including, or especially, in baby bottles. However, the conclusion could also argue in favour of a continuation of existing practices to allow BPA, notwithstanding the concerns voiced by individual Member States or EFSA. Furthermore, the European Commission has to respond to the temporary ban of BPA use in food packaging initiated by Denmark. Initially, this response was most likely to occur within the regulatory scope of the REACH Regulation. However, developments took another course which will be discussed later in this contribution.

First, let us turn to the background of the temporary ban and the responses that are possible. The Danish temporary ban invoked by the Danish People's Party on the basis of the Danish DTU food assessment involved a temporary precautionary measure regarding BPA in materials in contact with food for children under the age of three (infant feeding bottles, feeding cups and packaging for baby food). The legal basis was the precautionary principle. Irrespective of the EFSA review or any scientific evidence, the regulation of a substance requires a legal procedure. In case Member States maintain their bans and EFSA reaches the conclusion that there are no scientific grounds justifying the precautionary principle, the European Commission has three options: i) to sue the Member State before the European Court of Justice, which will take years; ii) initiate a proposal that would fulfil some or all of the Members State's request concerning the use of BPA. Any proposal aiming to create consensus between the Commission and Member States will need a legal basis, and its implementation would also take several years before the law can become effective; and iii) establish the proper legal basis in REACH, ensuring that the possible health effects for consumers are addressed.

However, there are also two other possible routes. In order for the substance to be restricted on the basis of the REACH Regulation, Article 8 must be invoked. The European Commission may opt for this but a Member State may also propose a restriction procedure after they have prepared a REACH Annex 15 dossier (substance evaluation).

Due to its 2008 classification as 'reprotoxic substance' (ECB 2008), category two in the Classification, Labelling and Packaging of chemicals (CLP) report, another trajectory is the Authorisation procedure under Article 7. This procedure requires: i) that the BPA is nominated for the candidate list of hazardous substances; ii) that it is selected as a priority according to Annex 14 (REACH);[25] and iii) that the EU issues a legal decision with regard to adjustment of Annex 14, thereby either banning or allowing the use of BPA on the EU market. Only once such a decision is in place may a possible restriction or ban become active. Such a trajectory would, probably, take years.

Self-Regulation by Industry

Amidst the scientific, societal and regulatory debates from 2005 onwards, industry responded and the majority of producers offer alternative 'BPA-free' products on the EU and US markets, prior to any regulatory restrictions in the EU.[26]

What is the rationale behind their *modus operandi*? Taking a closer look at their websites, product information, press releases and contacts (phone, email, interviews), apparently industry anticipates that consumer fears may prompt regulatory action. Moreover, marketing a BPA-free product not only suggests a risk-averse engagement with consumer safety from the side of industry (BFR 2010) but also provides them with a trade advantage, staying ahead of possible market regulation that is 'bound to happen' (#4). Not all producers provide BPA-free alternatives. Some continue to market their products while providing 'uncertainty information' (Van Asselt and Petersen 2003) on the scientific debates about possible health effects of BPA.

Difrax is an outstanding example. Uncertainty information incorporates the whole set of informed ideas and insights held by experts and scientists on the uncertainties without labelling them as simply suffering an absence of knowledge. Difrax has opened a BPA forum on their website, www.difrax.nl. They use(d) this platform to inform the consumer on the current controversy involving BPA by adding numerous threads with scientific studies and risk assessment reports. Moreover, the link also includes an interview with a leading Dutch toxicologist, explaining the risks and uncertainties that exist with regard to usage of BPA in consumer and children's products, what the possible alternatives are and whether they are desirable, even though comments are not always to the advantage of the baby bottle producer. Furthermore, there is a helpdesk and a FAQ area and a section that provides information on the development of one particular alternative and its (dis)advantages in usage, one such disadvantage being that the bottle may show signs of melting after having been heated several times in a microwave and another that the effects of alternatives have not been evaluated either (BFR 2010).

Therefore, industry has been active both in labelling BPA-free products and in providing uncertainty information in the absence of any EU regulatory obligation. After new developments in November 2010, leading up to a EU-wide ban of BPA in baby bottles by March 2011, companies such as Difrax have also been forced to adapt to selling BPA-free products only. There is a general fear within industry that banning BPA in baby bottles will have a 'spill-over effect' to more and more products such as toys, soda cans and other food items. Trade implications are difficult to overlook. Since self-regulation has provided industry with a marketing advantage, online polls and websites carry responses from the general public, seemingly convinced by the BPA-free marketing and expecting BPA-free products to be 'safe', whereas product details on why certain choices have been made and which alternative substances are used are missing.

BPA Ban in Baby Bottles

In October, the EU Commission announced its 'intention' to ban the use of BPA in the production of baby bottles. According to Health and Consumer Commissioner Dali no decision was reached but discussions with Member States, industry and EFSA had been ongoing.[27] What is striking is that the intention came in the wake of the expert report by EFSA early in October 2010 that shows a continuation of its current stance.

Following a detailed and comprehensive review of recent scientific literature and studies on the toxicity of bisphenol A at low doses, scientists on the EFSA CEF[1] conclude they could not identify any new evidence which would lead them to revise the current Tolerable Daily Intake for BPA of 0.05 mg/kg body weight set by EFSA in its 2006 Opinion and re-confirmed in its 2008 Opinion

EFSA 2010

In general, the Panel acknowledged that uncertainty remains as to possible effects and that further research is needed. The Commission interpreted this scientific Opinion as an indication that health effects cannot be ruled out and the restriction therefore remains in place.[28] This regulatory behaviour is in line with scenario II (see below), although no formal direction towards banning had been agreed upon between Member States.

However, months earlier than expected at that time, the decision was approved by a committee of national government experts. Commissioner Dalli explained that he had 'tried his luck' in a meeting of national experts coming together on 25 November 2010. The deadline for implementation of a region-wide ban was set by the EU Commission for March 2011.

Furthermore, Commissioner Dalli announced that the ban specifically targets the use of BPA in baby bottles since 'the EFSA advice had thrown up "areas of uncertainty" which meant infant exposure to the chemical should be minimised'.[29] He then added:

We are not planning to take further action because the advice from EFSA does not give us a basis to do so. Everything we do is based on scientific advice from EFSA. If you look at the Opinion there is currently no need to move further on BPA.[30]

No reference was made to the lack of knowledge of alternatives that will be used for BPA in baby bottles.

However, the debate continues. During the committee meeting, four national experts abstained, including the UK. The UK Food Standards Agency indicated that 'insufficient time' was given to examine the Commission's proposal and that it maintains its findings of there being not enough evidence that BPA poses a risk.[31] Scientists at the Medical Research Council's Human Reproductive Sciences Unit in Edinburgh and at the Department of Obstetrics and Gynaecology at Canada's McMaster University were among the first to respond that the Commission has acted with 'extreme caution'. Some scientists question whether this is perhaps a 'political' decision, rather than a 'scientific' one. Industry's first response was one of 'deep disappointment' over 'unconvincing' reasons behind the ban[32] and emphasised that EFSA had found 'no credible evidence' for altering current exposure levels and also that the Opinion was widely seen as an endorsement of the current levels of the substance in food contact materials.

The Food Standard Safety Agency in the UK called out to all stakeholders, including food packaging companies and trade bodies, for their views on the national implementation process of the EU Directive that outlawed the use of BPA in baby bottles from March 2011 onwards (Harrington 2011). The FSA remains hesitant about the need for the Directive banning the use of BPA in baby bottles but does not want to dismiss worries among a large group of people about BPA-containing feeding bottles.

Directive 2011/8 takes its legal basis in the precautionary principle and took effect in March 2011. It stipulates that, since June 2011, all BPA-containing infant feeding bottles should be replaced. Industry argues that this will take more time than legislation dictates and according to the North American Metal Packaging Alliance, the race is on to find alternatives. There are expected to be various alternatives and information on safety data is likely to not be available soon.

Despite the EU Commission's plans and the fact that BPA in baby bottles is now regulated, the scientific uncertainty of its possible risk remains and has not been addressed by the regulatory response. Scientific uncertainty remains and the alternatives now used in baby bottles may vary from manufacturer to manufacturer and are unknown to the consumer. Moreover, the characteristics of the replacement or substituting chemical substances in some cases remain less clear than in the case of BPA, which had been subject to several substance evaluations. The speed with which the decision was taken seems to suggest that a ban had apparently become the sole option. Nevertheless, the EU Commission in fact had several options, which we discuss by means of four scenarios.

Scenarios

Scenarios may provide an instrument through which possible futures may be explored in a structural way. According to Giddens (1991), the use of scenario analysis is intrinsic to reflexivity in the context of risk assessment and evaluation. And although scenarios are not widely applied in risk research (Van Asselt *et al.* 2010), it is recognised in the risk community that in cases of uncertainty the scenario process is one that explicitly accepts uncertainty in presenting alternative descriptions of the future (Rogers 2001).

It is important to emphasise that the use of scenarios is not a tool to predict the future but rather a tool that enables actors (regulators, scientists, industry and so forth) to gain insight into how existing uncertainties, in this case with regard to regulatory action, might play out in various futures (Van 't Klooster 2007; Van Asselt *et al.* 2010). In the first two scenarios the lead actor is the regulator, whereas industry takes the lead in the other two scenarios. In the former (actor-led) scenarios, the Opinions issued by EFSA or other advisory expert bodies do not determine the regulator's actions but could be seen as a stimulus.[33]

Scenario I: REACH Ban[34]

Industry anticipates that with the entry into force of REACH, restrictive measures against BPA are inevitable. NGOs working on BPA expect that the current nomination procedure of certain types of phthalates under REACH will set the precedent for BPA. This would imply that the authorisation procedure will apply to BPA. However, such a restriction would take years to enforce.

In this scenario, EFSA's Opinion would be crucial. Based on EFSA's scientific Opinion the European Commission and the Commission could request Denmark, Germany, France and Sweden to lift their moratoria and temporary restrictions as part of a risk management measure. These Member States, in co-operation with NGOs and Members of the European Parliament, would press for a ban via the candidate list. This will not succeed.

In this scenario, BPA would be registered in December 2010 and would be placed on the Candidate List under REACH. This procedure would lead to a ban, but after a lengthy procedure of a couple of years. On a short-term basis, the bans currently in place would be lifted and BPA would continue to be used. However, in the long term, the NGOs and Member States are confident to see this temporary permission being turned into a definite ban. In such a scenario the case would develop via regular procedures and would raise no related WTO issues. Once in place, BPA would no longer be used in the production of baby bottles and other children's products. The regulator is the lead actor and NGOs as well as Member States would only be moderately satisfied with this result.

Scenario II: Emergency Ban

In this scenario, the EFSA Opinion would serve as a stimulus. EFSA's Opinion will state that current practices can remain in place and will maintain that there is no scientific evidence of human health risk under current exposure limits. However, due to scientific uncertainty and the controversy that it has sparked, EFSA acknowledges some concern and will recommend altering the nature of the current exposure limit, the Tolerable Daily Intake (TDI) of 0.05 mg/kg per bodyweight per day, into a temporary TDI (in accordance with the July 2010 intermediate conclusion published on their website). Moreover, they see the need for further research. This advice implies that the Commission would lift the temporary bans and restrictions invoked by various Member States (Denmark, France, Germany and Sweden).

However, partly in response to political pressure from the Council, the EP and NGOs, the Commission decides to interpret EFSA's Opinion as an indication of a substantial risk. The Commission would provide the legal basis for restrictive regulation and it would continue to allow Member States to invoke the precautionary principle while preparing the official ban via the REACH authorisation procedure. In this way, BPA would be banned from the European market, first in fact and later in law.

In this scenario, the regulator would initiate a procedure that would lead to the ban of BPA. The substance would no longer be used in baby and children's products. NGOs would be highly satisfied as well as Member States, since they can maintain their bans. Although they are all initially pleased with the self-regulatory behaviour by industry, a new worry could develop over the use of alternatives in general and the lack of knowledge about its effect on human health in particular (compare Scenario IV).

Scenario III: Voluntary Renunciation

In this case, EFSA would conclude that – notwithstanding the presence of scientific uncertainty about consumer risks – there is not sufficient evidence to justify the current Danish and other moratoria. Furthermore, pragmatic trade concerns on the side of industry, DG Enterprise and within certain EU Member States, weigh heavily on the final decisions since BPA is used in a wide range of products. In this scenario, it is less likely that the regulatory footsteps of the phthalates case would be followed since there are some important distinctions. There are several types of phthalates and only particular phthalates are applied in the production process of children's products. Hence the use of these particular phthalates can be limited more easily without any effect on the position, production and application of other types of phthalates on the EU and global markets. With BPA this is not possible since it concerns only one substance that is applied to a wide range of products. Consequently restrictive measures would have an immense impact on the EU and global trade markets. However, industry continues to renounce the use of BPA in food packaging and children's products voluntarily in an attempt to show its willingness to compromise in this rather small area of production in exchange for continued use in other areas. This would allow an uncontroversial continuation of the use of BPA in a variety of other products such as bicycle helmets, electronic devices and in the medical field.

In this scenario, regulation seems to be lagging behind practice. The only option left for a regulator in such a case would be codification. Overall, in this scenario the substance would not be banned and use in baby and children's products would formally be allowed, although this hardly ever occurs in practice. Due to voluntary renunciation by industry, NGOs would be less concerned about this formal reality. The lack of a ban (or the prospect on a ban) worries the NGOs and Member States who would have to lift their bans, although self-regulation by industry eases these worries slightly.

Scenario IV: Uncertainty Information

In this scenario BPA would not be banned. EFSA's Opinion would recommend maintenance of the current exposure limits and would see no need for a ban. The Commission would compel Denmark, France and Germany to lift their bans.

Industry would continue to self-regulate the use of BPA in food packaging and children's products. However, the continued use of BPA in food packaging and baby bottles (a strategy that seemed to be only marginally applied in scenario III) would become dominant due to the practical impediments faced by industry in BPA-free production. Evaluations, such as the evaluation report by Health Canada, would continue to identify BPA traces in BPA-free products. Furthermore, the synthetic alternatives used in BPA-free products, such as polypropylene and polyether sulfone, cause too many problems. It is argued that polypropylene releases more substances into food than polycarbonate and faces consistent debate about exposure limits. Bottles made of polyether sulfone face a more fundamental objection because this starting material is used for the synthesis of plastic materials and has been even less the subject of scientific investigation than BPA.

Since the first headlines on problems relating to alternatives have already appeared, industry would realise that the problems have not been solved but, rather, they have shifted, and it would require a completely different approach to win over consumer trust. The approach in which the provision of uncertainty information takes centre stage would become the dominant strategy in communicating about BPA. All in all, BPA would not be banned and would be used widely in baby and children's products due to the problems that have surfaced with regard to alternatives. NGOs would consider this development as worrisome and those Member States that had been forced to lift their bans would not be prepared for this new situation in which the use of BPA would increase. A sensible strategy would have to be developed.

Conclusion

This chapter has articulated the controversy about BPA and has addressed its regulatory challenges within the EU. It aims to shed light on the regulatory complexities in uncertain risk dossiers. BPA can be best understood as an uncertain risk. Despite many risk assessments and evaluations of evidence, it remains unlikely that there will be any scientific conclusiveness in the near future or even the midterm. However, this case poses a serious regulatory challenge. Different scenarios are possible and plausible and have been outlined. These scenarios are not, and should not be used as, a tool to predict the future but rather as a tool to illustrate the dynamics of the case and its possible regulatory outcomes.

This chapter has explored possible regulatory futures. However, it also aims to suggest that the BPA regulatory controversy deserves academic attention. Despite the fact that BPA and the subsequent emergency ban is headline news, there are no explicit case descriptions in the academic literature. However, due to its complicated regulatory nature in which uncertainty takes centre stage, there is a definite need for review.

Notes

1 'BPA', 2010, available at: <http://www.bisphenol-a.org>.
2 Regulation 1907/2006 concerning the Registration, Evaluation, Authorisation and Restriction of Chemicals (REACH), OJ 2006 L 396/1, available online: <http://ec.europa.eu/environment/chemicals/reach/reach> (accessed on 15 August 2010).
3 See table 7.1: selection of headlines.
4 References made to interviews will be numbered (# 1, #2 etc.). The respondent's list is anonymous.
5 Directive 2005/84/EC amending Council Directive 76/769/EEC.
6 *Ibidem*, 24.
7 #3 Interview with members of Bureau REACH. #4 Interview with NGO representatives working on chemical substances in the EU.
8 #1 Interview with senior policy advisor and former Members of the ECHA Management Board. #5 Interview with expert risk assessment panel member Food and Safety Authority, the Netherlands.
9 #2 Interview with professors in chemical engineering at Dutch Universities.
10 *Ibidem*, 43.
11 Consistency assessment refers to whether a specific test can be repeated and is considered to be significant by EFSA to identify possible patterns that would underline the presence of a risk.
12 *Ibidem*, 120.
13 The studies they refer to include: Negishi *et al.* (2004); Carr *et al.* (2003).
14 Freshfield Bruckhaus Deringer describes the influence on China and the Asian region in their document *Food Safety Update: BPA*, whereas Majone sheds light on the consequences for the US (Majone 2010) and Bowman (Bowman and Van Calster 2007) looks at the global implications.
15 Regulation 1907/2006, *supra* note 2.
16 *Ibidem*, p. 50.
17 The term 'comitology' relates to 'the procedures by which the Commission makes decisions about the implementation of European Union (EU) legislation in collaboration with committees of national experts. These committees of national experts, usually civil servants from the national administrations, assist and advise the Commission on a wide variety of matters related to the implementation of individual pieces of Union legislation'. See Dogana (1997).
18 *Ibidem*, p. 50.
19 This does not only apply to the chemical sector, but on a more general note to regulatory regimes on risk: see for example Jasanoff (2005) and Versluis (2003).
20 The purpose of phthalates is to soften the polyvinyl chloride in products and facilitate the molecules to slide next to each other in the production process.
21 EU Commission IP/99/829, 1999; p. 2 available online: <http://europa.eu/rapid/pressReleasesAction.do?reference=IP/99/829&format=HTML&aged=1&language=EN&guiLanguage=en> (accessed on 4 July 2011).
22 Directive 2005/84/EC.
23 #1 Interview with former Members of the ECHA Management Board; #4 Interview with NGO representatives working on chemical substances in the EU.
24 #3 Interview with members of Bureau REACH; #4 Interview with NGO representatives working on chemical substances in the EU.
25 This procedure is in line with the current nomination of the various types of phthalates (DINP, DEHP and DBP) under REACH.
26 Only in the US are there any restrictions posed at state level (in California and Missouri). Canada is the first country worldwide that has imposed restrictions on the use and import of BPA.

27 Various news reports such as, 'EC in the UK, Chemical to be banned in the use of baby bottles', 2010, available online: <http://ec.europa.eu/unitedkingdom/press/frontpage/2010/1005_en.htm> (accessed on 20 May 2011).
28 Various news sources such as: *Food Production Daily, EU to ban BPA in baby bottles* (2010), available online: <http://www.foodproductiondaily.com/Quality-Safety/EU-to-ban-bisphenol-A-in-baby-bottles> (accessed on 20 May 2011).
29 *Ibidem*, p. 80.
30 *Ibidem*.
31 *Ibidem*, p. 78, 79 and 80.
32 *Ibidem*.
33 These scenarios were developed prior to this article and the release of the EFSA advice in September 2010. The EFSA advice has been released and the EU Commission has announced its plans to ban BPA by summer 2011. However, these scenarios are still useful in order to understand and define the situation that presented itself on 26 November 2010. Furthermore, it will help us to look into possible futures since the BPA ban is not in place yet.
34 With recent regulatory developments concerning BPA that occurred amidst revising this contribution, it became clear that Scenario I has not taken place and is no longer likely, whereas the other scenarios all have ingredients that cannot be excluded in future. Nevertheless, it was one possible option available to the European Commission prior to the decision favouring a ban. Thus, although this scenario ceases to be realistic and is at odds with the most recent developments, it is at the same time an illustration of the regulatory challenges that need to be confronted when dealing with uncertain risks.

References

Apfel, R.J. and Fisher, S.M. (1984) *To do no Harm: DES and the Dilemmas of Modern Medicine*, New Haven: Yale University Press.
Babyplaza (2010) 'Het gebruik van Bisphenol A in babyflessen', available online: <http://babyplaza.be/> (accessed 2 August 2010).
Bowman, D. and Van Calster, G. (2007) 'Reflecting on REACH: Global Implications of the European Union's Chemical Regulation', *Nanotechnology Law and Business*, 4(3): 375–84.
Bundesinstitut fuer Risikobewertung (BFR) (2010) 'BPA', available online: <http://www.bfr.bund.de/cd/template/index> (accessed 22 May 2011).
Carr, R., Bertasi, F., Betancourt, A. *et al.* (2003) 'Effect of Neonatal Rat Bisphenol A Exposure on Performance in the Morris Water Maze', *Journal of Toxicology and Environmental Health*, 66(21): 2077–88.
Danish Ministry of Food and Agriculture (2010) 'Danish Ban on BPA in materials in contact with food for children age 0–3', available online: <http://www.fvm.dk/Default.aspx?ID=18488&PID=168823&NewsID=6014> (accessed 2 August 2010).
Dodds, E.C. and Lawson, W. (1938) 'Oestrogenic Activity of Certain Synthetic Compounds', *Nature*, 139: 627.
Dogana, R. (1997) 'Comitology: Little Procedures with Big Implications', *West European Politics*, 20(3): 31–60.
EFSA (2006) 'Opinion of the Scientific Panel on Food Additives, Flavourings, Processing Aids and Materials in Contact with Food on a request from the Commission related to 2', *EFSA J.*, 2, available online: <http://www.efsa.europa.eu/en/scdocs/doc/428.pdf> (accessed 10 August 2010).

——(2010) *Bisphenol A*, available online: <http://www.efsa.europa.eu/en/ceftopics/topic/bisphenol.htm> (accessed 15 August 2011).

EPA (n.d.) 'Bisphenol A CASNR 80–05 2006–8', available online: <http://www.epa.gov/iris/subst/0356.htm> (accessed 2 August 2010).

Erickson, B.E. (2008) 'Bisphenol A under Scrutiny', *Chemical and Engineering News* (American Chemical Society), 86(22): 36–39.

European Chemicals Bureau (ECB) (2003) *European Union Risk Assessment Report. 4,4'-isopropylidenediphenol (Bisphenol-A)*, CAS No. 80-05-7 and 3rd Priority List (EUR 20843 EN), available online: <http://ecb.jrc.ec.europa.eu/esis/index.php?PGM=ora> (accessed 4 July 2011).

——(2008) *European Union Risk Assessment Report Draft: Updated Risk Assessment of 4,4'-isopropylidenediphenol (Bisphenol-A)*, CAS No. 80-05-7; EINECS No. 201-245-8, available online: <http://ecb.jrc.ec.europa.eu/esis/index.php?PGM=ora> (accessed 4 July 2011).

European Joint Research Commission (EJRC) (2010) 'Bisphenol A and Baby Bottles: Challenges and Perspectives', available online: <http://publications.jrc.ec.europa.eu/repository/bitstream/111111111/14221/1/eur%2024389_bpa%20%20baby%20bottles_chall%20%20persp%20(2).pdf> (accessed 15 August 2011).

FDA (2007) 'Bisphenol A and Phthalates', available online: <http://www.fda.gov/Food/FoodIngredientsPackaging/ucm166145.htm> (accessed 10 August 2010).

Ferguson, S.A. (2002) 'Effects on Brain and Behavior Caused by Developmental Exposure to Endocrine Disrupters with Estrogenic Effects', *Neurotoxicology and Teratology*, 24(1): 1–3.

Freshfields Bruckhaus Deringer (2010) *Food Safety Update: BPA*, available online: <http://www.fvm.dk/Default.aspx?ID=18488&PID=168823&NewsID=6014> (accessed 15 August 2011).

FSA (UK) (2010) 'Risk Assessment of BPA Update', available online: <http://www.food.gov.uk/foodlabelling/packagingbranch/foodcontactmaterialsbpa/> (accessed 14 August 2010).

Gezondheidsraad (2008) *Voorzorg met rede*, The Hague: Health Council of the Netherlands (Title translated as *Prudent Precaution*).

Giddens, A. (1991) *Modernity and Self-Identity: Self and Society in the Late Modern Age*, Cambridge: Cambridge Polity Press.

Goettlich, P. (n.d.) 'Various Contributions to Mindfully.Org, 2006–10', available online: <http://www.mindfully.org> (accessed 10 August 2010).

Gray, G.M., Cohen, J.T., Cunha, G. *et al.* (2004) 'Weight of the Evidence Evaluation of Low-dose Reproductive and Developmental Effects of Bisphenol A', *Human and Ecological Risk Assessment*, 10: 875–921.

Harremoës, P., Gee, D., MacGarvin, M. *et al.* (2002) *The Precautionary Principle in the 20th Century: Late Lessons from Early Warnings*, London: Earthscan.

Harrington, R. (2011) 'US Government Study Casts Significant Doubts over BPA Threat', in *Foodproduction Daily*, available online: <http://www.foodproductiondaily.com/Quality-Safety/US-Government-study-casts-significant-doubts-over-bisphenol-A-threat> (accessed 4 July 2011).

Health Canada (2008) 'Government of Canada takes action on another chemical of concern: BPA', available online: <http://www.hc-sc.gc.ca/ahc-asc/media/nr-cp/_2008/2008_59-eng.php > (accessed 2 August 2010).

Hermans, M. *et al.* (2012) 'Risk Governance', in S. Roeser (ed.), *Handbook of Risk Theory. Epistemology, Decision Theory, Ethics, and Social Implications of Risk*, Dordrecht: Springer, 1093–1117.

Heudorf, U., Mersch-Sundermann, V. and Angerer, J. (2007) 'Phthalates: Toxicology and Exposure', *International Journal of Hygiene and Environmental Health*, 210(5): 623–34.

Honma, S., Suzuki, A., Buchanan, D.L. *et al.* (2002) 'Low Dose Effect of in Utero Exposure to Bisphenol A and Diethylstilbestrol on Female Mouse Reproduction', *Reproductive Toxicology*, 16(2): 117–22.

International Risk Governance Council (IRGC) (2005) *White Paper on Risk Governance: Towards an Integrative Approach*, Geneva: IRGC.

Jasanoff, S. (2005) *Designs on Nature: Science and Democracy in Europe and the United States*, USA: Princeton University Press.

Kitzinger, J. (1999) 'Researching Risk and the Media', *Health, Risk and Society*, 1(1): 55–69.

Krayer von Kraus, M. (2005) *Uncertainty in Policy Relevant Sciences*, PhD thesis on file at the Technical University of Denmark.

Kubwabo, C., Kosarac, I., Stewart, B. *et al.* (2009) 'Migration of BPA from Plastic Baby Bottles, Baby Bottle Liners and Reusable Polycarbonate Drinking Bottles', *Food Additives and Contaminants*, 26(6): 928–37.

Lamb, J.C. (2002) 'Why You Should Ignore the Babybottle Scare', available online: <www.quackwatch.org/04ConsumerEducation/babybottle.html> (accessed 4 July 2011).

Löfstedt, R.E. (2005) *Risk Management in Post-trust Societies*, Basingstoke: Palgrave Macmillan.

Lozito, W. (2007) 'Product Naming: Is BPA Free the New Buzzword for Babies', available online: <http://www.namedevelopment.com/blog/archives/2007/04/product_naming_5.html> (accessed 2 August 2010).

Majone, G. (2010) 'Foundations of Risk Regulation: Science, Decision-making, Policy Learning and Institutional Reform', *European Journal of Risk Regulation*, 1(1): 5–19.

Meyers, R. (1983) *D.E.S., The Bitter Pill*, New York: Seaview/Putnam.

Mountfort, K., Kelly, J., Jickells, S.M. *et al.* (1997) 'Investigations into the Potential Degradation of Polycarbonate Baby Bottles During Sterilization with Consequent Release of BPA', *Food Additives and Contaminants*, 14(6–7): 737–40.

Murray, T.J., Maffini, M.V., Ucci, A.A. *et al.* (2007) 'Induction of Mammary Gland Ductal Hyperplasias and Carcinoma in situ Following Fetal Bisphenol A Exposure', *Reproductive Toxicology*, 23(3): 383–90.

National Toxicology Programme (2007) *The Evaluation of Risks to Human Reproduction 2007*, Expert panel evaluation of BPA, avaliable online: <http://ntp.niehs.nih.gov/> (accessed 2 August 2010).

Negishi, T., Kawasaki, K., Suzaki, S. *et al.* (2004) 'Behavioral Alterations in Response to Fear-provoking Stimuli and Tranylcypromine Induced by Perinatal Exposure to Bisphenol A and Nonylphenol in Male Rats', *Environ. Health Perspect*, 112(11): 1159–64.

Pidgeon, N., Kasperson, R.E. and Slovic, P. (2003) *The Social Amplification of Risk*, Cambridge: Cambridge University Press.

Renn, O. and Walker, K. (2008) *Global Risk Governance: Concept and Practice Using the IRGC Framework*, Dordrecht: Springer.

Rogers, M.D. (2001), 'Scientific and Technological Uncertainty: the Precautionary Principle, Scenarios and Risk Management', *Journal of Risk Research*, 4(1): 1–15.

Safe, S.H., Pallaroni, L., Yoon, K. *et al.* (2002) 'Problems for Risk Assessment of Endocrine Active Estrogenic Compounds', *Environmental Health Perspectives*, 110(6): 925–29.

Sathyanarayana, S., Karr, C.J., Lozano, P. *et al.* (2008) 'Baby Care Products: Possible Sources of Infant Phthalate Exposure', *Pediatrics*, 121(2): 260–68, available online:

<http://pediatrics.aappublications.org/cgi/content/full/121/2/e260> (accessed 22 June 2010).

Van Asselt, M.B.A. (2000) *Perspectives on Uncertainty and Risk: The PRIMA Approach to Decision Support*, Dordrecht: Kluwer Academic Publishers.

Van Asselt, M.B.A. and Petersen, A.P. (2003) *Not Afraid of Uncertainty (in Dutch)*, The Hague: Lemma/RMNO.

Van Asselt, M.B.A. and Renn, O. (2011) 'Risk Governance', *Journal of Risk Research*, Special issue 'Uncertainty, Precaution and Risk Governance', 14(4): 431–49.

Van Asselt, M.B.A and Vos, E. (2006) 'The Precautionary Principle and the Uncertainty Paradox', *Journal of Risk Research*, 9(4): 313–36.

Van Asselt, M.B.A., Van 't Klooster, S.A., Van Notten, P.W.F *et al.* (2010) *Foresight in Action. Developing Policy Oriented Scenarios*, London: Earthscan.

Vandenberg, L.N., Hauser, R., Marcus, M. *et al.* (2007) 'Human Exposure to BPA', *Reproductive Toxicology*, 24(2): 139–77.

Van 't Klooster, S.A. (2007), *Toekomstverkenning: Ambities en de praktijk*, (PhD Thesis on file at Maastricht University).

Versluis, E. (2003) *Enforcement Matters. Enforcement and Compliance of the European Directives in Four Member States*, Delft: Eburon.

Versluis, E., Van Asselt, M.B.A, Fox, T. *et al.* (2010) 'The EU Seveso Regime in Practice: From Uncertainty Blindness to Uncertainty Tolerance', *Journal of Hazardous Materials*, 184(1–3): 627–31.

Vogel, S.A. (2009) 'The Politics of Plastics: the Making and Unmaking of BPA Safety', *American Journal of Public Health*, 99(3): 559–66.

Vom Saal, F. and Hughes, C. (2005) 'An Extensive New Literature Concerning Low-dose Effects of Bisphenol A Shows the Need for a New Risk Assessment', *Environmental Health Perspective*, 113(8): 926–33.

Vom Saal, F. and Myers, J.P. (2008) 'Bisphenol A and Risk of Metabolic Disorders', *JAMA*, 300(11): 1353–55.

Vom Saal, F., Nagel, S.C., Timms, B.G. *et al.* (2005) 'Implications for Human Health of the Extensive Bisphenol A Literature Showing Adverse Effects at Low Doses: A Response to Attempts to Mislead the Public', *Toxicology*, 212(2–3): 244–52.

VWA (2006) 'Bisphenol A in babyflesjes', available online: <http://www.vwa.nl/onderwerpen/levensmiddelen-food/dossier/babyvoeding/nieuwsoverzicht/nieuws bericht/23763/geen-aantoonbare-afgifte-van-bisfenol-a-in-babyflesjes> (accessed 10 August 2010).

Ward Barber, J. (2010) 'Why the Food Industry is Fighting for BPA', available online: <http://www.theatlantic.com/science/archive/2010/05/why-the-food-industry-is-fighting-for-bpa/56098> (accessed 2 August 2010).

Wiener, J.B., Rogers, M.D. and Sand, P.H. (2010) *The Reality of Precaution: Comparing Risk Regulation the US and EU*, Herndon (VA): Stylus Publishing USA.

Witorsch, R.J. (2002) 'Low-dose in Utero Effects of Xenoestrogens in Mice and their Relevance to Humans: an Analytical Review of the Literature', *Food and Chemical Toxicology*, 40(7): 905–12.

WRR (2008), *Onzekere Veiligheid. Verantwoordelijkheden rond fysieke veiligheid*, Amsterdam: Amsterdam University Press (available in English under the title: *Uncertain Safety*).

Wynne, B. (2001) 'Creating Public Alienation: Expert Cultures of Risk and Ethics on GMOs', *Science as Culture*, 10(4): 445–81.

Part III

Taking Stock of Policy Fashions

8 Agencies as Risk Managers?

Exploring the Role of EU Agencies in Authorisation Procedures

Jinhee Kim, Christoph Klika and Esther Versluis

Introduction

While innovative technologies promise to act as a motor for economic progress as well as societal benefits, they also invoke increasingly complex questions regarding institutional arrangements to govern their accompanying uncertainties and risks (Wynne 2001; Jasanoff 2005; Löfstedt 2005; Van Asselt and Vos 2006, 2008). How can we stimulate nanotechnology to produce smaller, lighter, faster and/or energy-efficient devices with more or new functionalities, while at the same time ensuring that the possibility of long-term health effects and/or environmental impacts is adequately accommodated? How can we regulate the chemical Bisphenol A (used in the production of baby bottles and teethers for example) for which the risk assessments are contradictory and thus uncertain regarding possible adverse health effects for babies (see also the chapter by Fox *et al.* in this volume)?

In line with the international process of 'agencification' (e.g. Hutter and Power 2005; Braun and Gilardi 2006), there is a noticeable trend in the EU to delegate specific tasks of risk regulation – here defined as governmental interference with market or social processes to control potential adverse consequences to health, safety and the environment (Hood *et al.* 2002) – to agencies. This trend of delegation is rooted in the belief that agencies – through (independent) expertise and apolitical, high quality evaluations – will produce better decisions, more efficiency and increased accountability, particularly in areas of high technical complexity (e.g. CEC 2002; Majone 2002). The creation of the European Food Safety Authority in 2002 can be seen in this light. The assumed 'mismanagement' of the BSE crisis by the European Commission triggered the call for a 'regulatory model that was more credible on output-legitimation grounds; that is, with a more 'independent' scientific basis of risk assessment (Skogstad 2003: 329–30). This illustrates the underlying assumption that independent scientific expertise, detached from political influence, will bring about better regulation of risks. It is believed that expertise accumulated by agencies and manifested in sound risk assessment will enhance the quality of risk management. Agencies thus seem to be a response to the ideational separation between risk assessment and risk management.

Risk assessment hereby refers to the purely scientific and technical assessment of risks, which will serve as a basis for decision making on risks. It is performed by scientists and experts who have knowledge and expertise of the risks in question. Risk management then refers to the policy decisions that ultimately aim at regulating the risks. On the basis of information offered by risk assessors, risk managers determine what should be done to control the risks. Of the more than 30 EU agencies, a considerable number implicitly or explicitly deal with risks such as pharmaceuticals, food and chemical safety, occupational health or aviation.[1] The most well-known, and most analysed, are the European Medicines Agency (EMA) providing advice on the authorisation of certain pharmaceuticals, and the European Food Safety Authority (EFSA), providing, amongst others, advice on the authorisation of genetically modified food and feed. In both cases, the agency is responsible for risk assessment while the European Commission, in co-operation with the Member States via the comitology procedure, is responsible for risk management by taking the final decision regarding authorisation.

While theoretically it may be possible – and conceptually perhaps even desirable – to separate risk assessment from risk management, several scholars argue that it is difficult in practice, and counterproductive even, to completely separate these two activities (Jasanoff 2005; Millstone and Van Zwanenberg 2001; Majone 2010). While conceptually distinct, there exists an unclear segregation between risk assessment and risk management in practice. Accordingly it may come as no surprise that several scholars analysing the working and functioning of EMA and EFSA conclude that these agencies are the *de facto* risk managers, as they seem to play a crucial role in the actual decision making of the authorisation of pharmaceuticals and genetically modified food and feed (Van Asselt and Vos 2008; Borrás 2006; Gehring and Krapohl 2007; Vos 2003).

In this contribution we take this perception of EU agencies as the *de facto* risk managers as the starting point for our analysis. Based on literature analysis, we identify how and under what conditions EMA and EFSA qualify as *de facto* risk managers. Given that EU agencies are increasingly lined up to take up similar responsibilities as those undertaken by EMA and EFSA, the understanding of the role of EU agencies in risk regulation bears considerable relevance. In terms of authorisation procedures, the recently established European Chemicals Agency (ECHA) is the most prominent example. The overarching aim of this chapter is to learn from the experiences of EMA and EFSA with regard to authorisation procedures. The main issue at stake is what these 'older' cases tell us about the role ECHA is likely to play in the authorisation of chemical substances. Can we – from the experiences with EMA and EFSA – distil certain circumstances or conditions that are likely to 'predict' whether ECHA will in turn also become the *de facto* risk manager?

EMA, EFSA and Authorisation Procedures

The European Medicines Agency (formerly the European Agency for the Evaluation of Medicinal Products) was established in 1993 (Regulation EEC No. 2309/93), and is one of the oldest EU agencies dealing with risks. The founding regulation of EMA was amended in 2004 (Regulation 726/2004), introducing the 'compulsory' centralised authorisation procedure for all medicinal products for human and veterinary use derived from biotechnology or high-technology process and products intended for the treatment of HIV/Aids, diabetes, cancer or other immune dysfunctions. EFSA was created after the BSE crisis in order to 'reinforce the present system of scientific and technical support which is no longer able to respond to increasing demands on it' (Preamble to Regulation 178/2002 at recital 33). Since April 2004, GM food and feed, or food and feed products, containing or consisting of GMOs have been regulated under the centralised authorisation procedure (Regulation 1829/2003). While Table 8.1 lists considerable differences between both agencies in terms of size and budget, the authorisations of both pharmaceutical products and GMOs go through relatively similar procedures in order to be marketed in the EU (see Figure 8.1).

EMA and the Authorisation of Pharmaceuticals

Recital 19 of the Preamble to the revised founding regulation (Regulation 726/2004) makes clear that EMA is to be seen as the risk assessor. It states that 'the chief task of the Agency should be to provide Community institutions and Member States with the best possible scientific opinions so as to enable them to exercise the powers regarding the authorisation and supervision of medicinal products'. In addition, Article 57 of this Regulation states that EMA provides the Member States and the EU institutions scientific advice on any question relating to the evaluation of the quality, safety and efficacy of medicinal products. This risk assessment is arranged via six Scientific Committees in EMA.[2] Among them, the Committee for Medicinal Products for Human Use (CHMP) and the Committee for Medicinal Products for Veterinary Use (CVMP) are responsible for issuing scientific opinions after their risk assessments. Both committees are comprised of

Table 8.1 Overview of EMA and EFSA

	EMA	EFSA
Year of establishment	1993	2002
Location	London, UK	Parma, Italy
Budget for 2012 (M€)	222.5	78.8
Number of Staff	866 (in 2011)	494 (in 2011)
Number of Scientific Committees	6 Scientific Committees	1 Scientific Committee and 10 Scientific Panels

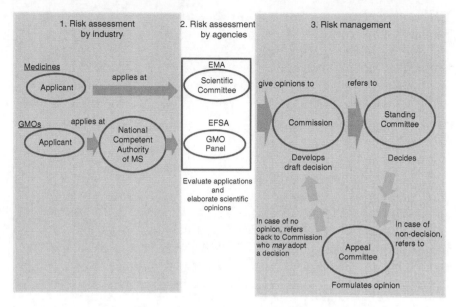

Figure 8.1 Centralised authorisation procedures: comparing EMA and EFSA

scientific experts acting as representatives from the National Competent Authorities (NCAs) in the field of pharmaceutical regulation.

The formal procedure for the centralised authorisation begins when an application is submitted by pharmaceutical companies to EMA. However, companies are encouraged to submit a 'pre-submission request form (Eligibility)' at the earliest 18 months or at the latest seven months, prior to submission of application. The EMA confirms the eligibility of the centralised procedure and assigns a Rapporteur and usually also a Co-Rapporteur. After this step, applications are submitted together with administrative data including information on the product, the manufacture (including packaging) and batch testing, and final batch release sites. More importantly, expert reports should also be submitted by applicants. Expert reports include information on chemical, biological, pharmaceutical, toxicological and clinical tests. In addition, a pharmaco-vigilance plan – describing in detail a risk management plan for the post-authorisation period – should be provided to EMA. It implies that by conducting all relevant tests of medicinal products and providing necessary information on risks to the EMA, pharmaceutical companies actually take part in risk assessment of the authorisation procedure.

In the case of medicines for human use, the CHMP conducts the risk assessment. It has to deliver an opinion within 210 days. Rapporteurs carry out risk assessments with their assessment team at their original national competent authority. The process is relatively interactive as questions and issues are raised back and forth between the assessors and the applicant. The Rapporteurs evaluate the information submitted by the applicant, and make the final assessment report. On or before day 210, the CHMP adopts its opinion. Adopting opinions by consensus is the usual practice, yet adoption by majority occurs as well, in which case the names of the dissenting Committee members must be mentioned in the opinion. The EMA has 15 days to transmit its opinion to the Commission, and this officially concludes the risk assessment by the EMA.

A detailed look at this procedure reveals that the risk assessment is actually shared by three actors: the applicants, the EMA and NCAs. Initial tests of medicinal products are conducted by pharmaceutical companies that develop new medicines, and the EMA carries out risk assessments based on the data and information provided by the industrial applicant. There is regular and frequent interaction between CHMP members – particularly the Rapporteurs – and the applicants. The involvement of the NCAs is also crucial. As all scientific experts in the EMA stem from about 40 national competent authorities, and Rapporteurs make use of national apparatus when assessing applications, it is only fair to argue that these actors play a crucial role in the risk assessment. In fact, there is very well-organised and continuous contact between EMA and NCAs, particularly through Committee meetings during the risk assessment period.[3] Rapporteurs and Co-Rapporteurs regularly (sometimes on a weekly basis) contact each other, CHMP members meet face-to-face for three to six days per month for Committee meetings, and have on average three to four teleconferences and one web-based meeting per month.[4] Especially when interactions via email are counted, CHMP members remain in contact with other CHMP members from different national competent authorities on a daily basis.[5] In sum, the EMA does not carry out its risk assessment in isolation. In practice, risk assessment for the pharmaceutical authorisation is carried out by the 'tripartite' of assessors.

Risk management, on paper, begins when the EMA's opinion is transmitted to the Commission.[6] Within 67 days, including 15 days of transmission of the opinion from EMA, the Commission needs to deliver a decision. Since the EMA's opinions are not legally binding, the Commission is not obliged to follow them. However, if the Commission's proposal is not in accordance with the opinion of EMA, the Commission needs to provide a detailed explanation of the reasons for the deviation (Article 10, Regulation 726/2004). Moreover, in the case of scientific or technical doubt, it can neither amend nor ignore the opinion of the EMA but must refer the matter back to the agency where it is examined again by the Scientific Committee (Gehring and Krapohl 2007).

By forwarding the proposal to the Standing Committee on Medicinal Products for Human Use (the relevant comitology committee), the Commission starts the 22-day Standing Committee (written) consultation, addressing 'only legal and public health matters (which means in principle no further linguistic review)' (EMA 2008: 4). Thus, if no Member State raises any written objections or questions about the proposal within 22 days, the Standing Committee is deemed to agree with the proposal. Then, within 15 days, the

Commission issues a decision to grant the marketing authorisation for five years.

While on paper the Commission and the Standing Committee are the risk managers, reality shows that the EMA is the dominant actor here. Although the Commission is officially not bound by EMA opinions, it has not yet deviated from the content of EMA opinions.[7] Similarly, Gehring and Krapohl (2007) argue that the Commission does not make use of a separate scientific apparatus to scrutinise EMA opinions. The recent evaluation report of the EMA presented to the Commission states that opinions from the Scientific Committees of the EMA are 'directly followed by a European Commission decision' and that 'even though all committees have a scientific advisory function, it is considered that, in practice, the outputs of these "opinion-making" committees are more binding, and have more political impact' (Ernst & Young et Associés 2010: 193).

In sum, we seem to be dealing with a risk regulatory regime in which the risk assessment is at least jointly conducted by the agency, industrial applicants, and NCAs, and in which the risk management phase comprises a Commission who rubber-stamps opinions from the EMA when authorising medicinal products.

EFSA and the Authorisation of GMOs

Similar to the EMA, EFSA's founding regulation clearly defines its main task in risk assessment. It states that EFSA provides 'a comprehensive independent scientific view of the safety and other aspects of the whole food and feed supply chains' and that 'scientific advice and scientific and technical support on human nutrition in relation to Community legislation and assistance to the Commission' (Preamble to Regulation 178/2002 at recital 36). A clear demarcation between risk assessment and risk management becomes even more apparent on EFSA's website. It states that the role of EFSA is 'to assess and communicate on all risks associated with the food chain', and that 'EFSA's advice serves to inform the policies and decisions of risk manager'.[8] As in the case of pharmaceutical products, a suitable Scientific Panel within EFSA is chosen to conduct the risk assessment; in the case of GMO authorisation this is the GMO Panel.

The official authorisation procedure begins when an application is submitted. Applications are not directly submitted to EFSA but to the NCA of the EU Member State where the product is first to be placed on the market. At this point the NCA functions as a 'post-box' and does not have to assess the dossier but 'simply' forwards the application to EFSA.[9] The application must include information on the method of production and manufacturing, a copy of the studies demonstrating that the product does not cause any adverse effects on health or environment, an analysis and a statement ensuring the safety and no ethical or religious concerns, and a proposal for labelling the food. Furthermore, a complete technical dossier and a monitoring plan for environmental effects are required for applications regarding GMOs or food containing or consisting of GMOs (Regulation 1829/2003, Article 5).

In contrast to the EMA, there is no pre-submission consultation between applicants and EFSA. Once an application is validated, the GMO Panel within EFSA conducts the risk assessment of the safety of GM food or feed for humans, animals and the environment up to a time limit of six months. The Panel currently has 21 members who are appointed through an open selection procedure on the basis of their proven scientific excellence. Therefore, they are not representatives from the Member States but experts independent of the NCAs and responsible for the regulation of food stuffs. Similar to the EMA however, Rapporteurs are appointed among members of the Panel and take the leading role in risk assessment. If necessary, EFSA may ask the appropriate food assessment body of a Member State to carry out a safety assessment. When the risk assessment is finalised, the GMO Panel adopts its opinion in a plenary meeting. While we know from the chapter by Navah *et al.* in this volume that there are huge differences between Member States in the level of acceptance of GM products, it is striking that since the beginning of its work, the Panel has always adopted 'positive' opinions on GMOs by unanimity.[10]

As is the case for the EMA, the GMO Panel does not carry out its own studies during the risk assessment process. Rather, the Panel makes a full risk assessment based on the data received from the applicants. Indeed, due to the important role of industry as a (co-)risk assessor, some scholars argue that EFSA's risk assessments amount to *meta-reviewing* of applicants' assessments instead of an *independent examination* of them (Van Asselt *et al.* 2009).

Since the Member States are not directly represented in the GMO Panel – it being staffed by independent experts – the role of NCAs during the risk assessment period is not as strong as would be the case for pharmaceutical authorisations. As a consequence, EFSA only occasionally interacts with NCAs when they attend Panel meetings as guests.[11] In the case of the GMO Panel, there are on average only four official meetings organised for NCAs per year.[12] Therefore, to some extent the GMOs authorisation involves the tripartite risk assessors – EFSA, industrial applicants and, to a lesser extent, NCAs.

During the risk management phase the Commission has three months to formulate a proposal after receiving an opinion from EFSA. As in the case of the EMA, the Commission is not obliged to follow the agency's opinions but all Commission proposals favouring authorisation have been based on positive opinions from EFSA. The proposal is then submitted to the relevant comitology committee: the Standing Committee on the Food Chain and Animal Health (SCFCAH). The SCFCAH has a time limit of three months to adopt the proposal by a qualified majority vote.

While we see a unique level of consensus within EFSA and the parent DG Health and Consumers favouring the authorisation of GM products, this certainly is different in the comitology committee. The SCFCAH has never been able to reach an agreement on Commission proposals, and thus all cases of GMOs authorisations had to be referred to the Council of Ministers.[13] Here the striking disagreement continues and the Council also failed to adopt any decision by a qualified majority vote. This inability of the standing committee and the Council to reach a decision is mainly due to disagreement among the Member States on the decisions about the safety of GMOs for humans and the environment (see also Navah *et al.* in this volume). The authorisation of GMOs is a highly politicised issue which brings about contradicting positions, and Member State representatives being known for voting based strictly on instructions and mandate from their ministries (Vos and Wendler 2006) certainly do not help to overcome this deadlock.

Due to the inability of the standing committee and the Council to achieve the required qualified majority, proposals are referred back to the Commission for a final decision. Before the new comitology rules entered into force, this principally implied that the Commission had an important role to play in the authorisation of GMOs. Notably, in all cases it has been up to the Commission to grant or reject the authorisation of GMOs in the EU. Despite the fact that there is a simple majority of the Member States opposing the authorisation in both the comitology committee and the Council, the Commission has always authorised GMOs in line with opinions of EFSA. As seen in the authorisation of pharmaceuticals, a similar pattern of the Commission's behaviour is observed. When making a final decision about the authorisation, the Commission rubber-stamps EFSA's opinions (see, for example, Chalmers 2005). It has to be seen, however, whether the new comitology rules – by which the Commission may, and is no longer obliged to, adopt a final decision in case of disagreement in the standing committee and in the appeal committee – will change this current practice.

Comparison: EMA and EFSA as Risk Managers?

Risk Assessment: Tripartite Assessors but Different Composition of Scientific Committees/Panels

Although created with the explicit task of acting as the Union's risk assessor, both the EMA and EFSA share their risk assessment function with other actors. Applicants and NCAs play a crucial role in the risk assessment. Although the agencies complete the risk assessment of each application and present the results to the Commission, the role played by industrial applicants should not be disregarded. They conduct their own studies to prove the safety of their products and provide agencies with all relevant information and data to proceed with risk assessment. Likewise, agencies rely on the apparatus of NCAs when conducting risk assessment, particularly in the case of pharmaceutical authorisations. The explanation for more intensive NCA involvement in the case of the EMA seems to lie in the composition of the scientific committees/panels. While the CHMP is composed of representatives of

NCAs, independent experts form the GMO Panel. The historical context of agency creation underlines this difference. While the EMA was founded in response to the failure of the doctrine of mutual recognition in the common market of medicinal products, EFSA was largely created to restore public trust in the EU institutions to manage food safety after a severe crisis in governance at EU level. Thus, it is understandable that more emphasis was placed on harmonisation among national authorities dealing with medicines in the EMA, while in the case of EFSA the strategy was to build scientific credibility and public confidence through 'independent' expert advice (Levidow and Carr 2007). Consequently, representatives from NCAs are appointed in committees in the EMA, yet independent experts are recruited for the scientific panels in EFSA through an open competition.

Risk Management: Agencies as Managers but under Different Political and Procedural Situations

Both the EMA and EFSA heavily influence the risk management phase as the final decisions on authorisations of pharmaceuticals and GMOs by the Commission seem to suggest a rubber-stamping of the opinions of the two agencies. In other words, the Commission – which drafts the content of regulatory decisions – heavily relies on the agencies' opinions. However, the final decisions are made at different points in the authorisation procedure. In the case of pharmaceutical authorisations, the outcome of comitology procedures has always agreed with the drafts from the Commission. In the case of GMO authorisations, however, the Commission has always made the final decision after both the comitology procedures and the Council failed to reach any decision. Whether this procedure will last under the new comitology rules remains to be seen. The question is whether the Commission would be willing to oppose the minority group of the Member States against GMO authorisation in comitology and in the appeal committee, when it is no longer obliged to take the final decision. It may well be that the new comitology rules will cause more deadlock in GMO authorisation decision making.

While thus far the Commission has finally approved authorisation following the agency's opinion in both cases – thus making the agencies the *de facto* risk managers – the 'route' to the adopted decision differs. From a political perspective, we can argue that the Member States seek to reflect their opinions whatever the form of the authorisation procedure. Although the EMA as an EU agency is to be independent from any political influence, the composition of the CHMP ensures that the Member States have their voice heard and that different national opinions are well harmonised in the EMA. Even when an opinion is adopted by majority it is a common procedure to indicate minority opinions in the final opinion submitted to the Commission. Therefore, opinions from the EMA are difficult to challenge in comitology procedures 'since its members belong to the same national competent authorities as those of the agency's expert committee, and they also need a qualified majority to overturn such decisions' (Sabel and Zeitlin 2008: 286). It is the same in the Council because 'their own expert administrations are closely involved in their elaboration' (Gehring and Krapohl 2007: 215).

By contrast, the Member States have no official seat in EFSA's scientific panels, nor are they obliged to accommodate diverse national opinions. In addition, GMOs remain controversial for the European public. Citizens' concerns over GMOs are revealed in the *Eurobarometer* reports, indicating that 'overall Europeans think that GMO food should not be encouraged; and GM food is widely seen as not being helpful, as morally unacceptable and as a risk for society'.[14] Thus, it is difficult for Member State representatives to formulate a decision in comitology procedures and in the Council. Accordingly, the strongly divided Member States (see also Navah *et al.* in this volume) are unable to reach an agreement through comitology or in the Council (and expectedly in the new appeal committee), leaving the floor open to the Commission, which – always following EFSA's opinion – authorises the GM products.

A different route to authorisation in EMA and EFSA can also be analysed from a more procedural level. The time line in Figure 8.2 indicates that the

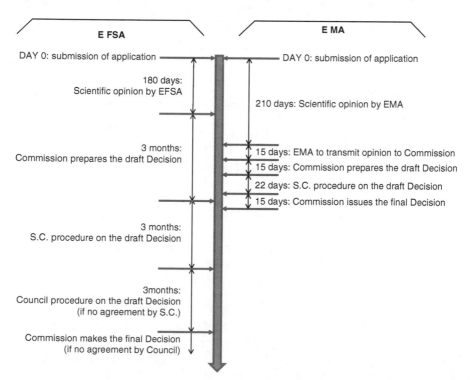

Figure 8.2 Timeline dimension of the centralised authorisation

decision-making bodies and procedures (Commission and comitology) in the pharmaceutical authorisation are under time constraints. The total number of days allowed for decision makers to reach a decision after the EMA issues an opinion is 67 days. However, it takes at least six months – and normally more than nine months (as the Council/appeal committee is involved) – to decide about GMO authorisation. One could question whether in cases of pharmaceutical authorisations, the Commission really has sufficient time to conduct a careful analysis of the risk assessment provided by the agency. Furthermore, decision-making in comitology is based on the written procedure in the pharmaceutical authorisation, while it follows the meeting procedure in the GMO authorisation. Under the written procedure, if at least one Member State in comitology raises concerns in writing on the Commission's draft, the matter has to be discussed in a plenary meeting. The draft can then only be rejected when the concerned Member State has convincing reasons to change the position of a majority of the Member States not having initially objected to the draft. Since this task is burdensome, it is more difficult under comitology procedures to deviate from EMA's opinions under the written procedure (see also Krapohl 2004).

To summarise, both agencies hold a leading position in risk management but under different premises: the Member States have a stronger role in the formulation of the EMA's opinion than is the case in EFSA, and the Commission has less time to decide on pharmaceutical authorisations. This seems to result in a smooth decision-making process without disagreement in the pharmaceutical authorisation, while the Member States are unable to reach an agreement in comitology procedures and in the Council/appeal committee in the case of GMOs. To put it differently, in the case of EFSA, agency influence in risk management is 'gained' because, due to the composition of independent experts in the GMO Panel, those nations opposing GMO authorisation cannot fully voice their position in EFSA opinions. Under comitology procedures, diverse national opinions can be heard but it is hard to negotiate or harmonise them since Member State representatives are bound to vote based on instructions and mandates from their ministries, as mentioned earlier.

Were the Member States to be better able to reach an agreement through comitology procedures or in the Council/appeal committee, it might become more common for them to overrule EFSA opinions, and thus downplay the role of EFSA in risk management. We could even conclude that agencies are only 'allowed' to play a role in risk management when the Member States feel that they have considerable control over the drafting of the agency's opinions. Furthermore, agency influence in risk management can be enhanced if procedural rules 'unofficially' constrain the freedom of decision makers to easily deviate from the agency's opinions. As stated, the new comitology rules might make EFSA less of a *de facto* risk manager as it could be anticipated that the Commission will no longer rubber-stamp GMO authorisations against the will of a minority group of the Member States when it is no longer obliged to decide.

Implications for the Authorisation of Chemical Substances: The Role of ECHA in Risk Regulation

While we, on paper, see a functional separation between risk assessment by European agencies and risk management by the EU institutions with regard to the authorisation of pharmaceuticals and GMOs, the previous section has illustrated that both EMA and EFSA seem to play an important role in risk management. Based on the insights gained from the EMA and EFSA scenarios, we now turn to the European Chemicals Agency (ECHA) and its (future) role in the authorisation of chemicals. ECHA was created in 2006 as part of the REACH regulation (Regulation 1907/2006) after a lengthy and heavily disputed legislative process (Pesendorfer 2006; Selin 2007; Lindgren and Persson 2008). The Regulation aims to ensure 'a high level of protection of human health and the environment' regarding the risks stemming from chemical substances, and at the same time ensures the competitiveness of the EU chemicals industry by way of a free circulation of substances on the Internal Market (Regulation 1907/2006, Article 1). As the title suggests, the REACH Regulation consists of various regulatory procedures, namely the Registration, Evaluation and Authorisation of Chemical substances (hence the acronym REACH).

From an institutional perspective, the authorisation procedure for chemical substances resembles those for pharmaceuticals and GMOs. The ECHA is to conduct a risk assessment via an internal committee, and to form an opinion with regard to risk management. After receiving an opinion from the agency, the Commission drafts a proposal for a decision to be taken in comitology. However, due to the path dependencies of EU chemicals policy, there are important differences that set ECHA apart from the other two agencies.

As a result of the EU's patchy regulatory framework prior to REACH, substances marketed before 1981 – the so-called 'existing substances' – were largely unregulated whereas substances marketed after 1981 – the so-called 'new substances' – were subject to a notification requirement involving extensive testing in order to obtain information about potential risks (Foth and Hayes 2008). One important rationale for the adoption of REACH was therefore to close the regulatory gap between existing and new substances. One regulatory instrument to achieve this is the 'registration requirement' comprising all substances currently or prospectively marketed; thus should a substance not

Table 8.2 Overview of ECHA

	ECHA
Established Year	2006
Location	Helsinki, Finland
Budget for 2012 (M€)	102.7
Number of Staff	472 (in 2011)
Number of Committees	3 Committees

be properly registered it cannot be marketed. Another instrument is the authorisation of the so-called Substances of Very High Concern (SVHC) (see Title VII REACH).[15] While EMA and EFSA are concerned with the authorisation of newly developed pharmaceuticals or GMOs, the authorisation process in ECHA is geared towards existing substances, and this bears important repercussions for the process of risk regulation.

ECHA in Risk Assessment

While the risk assessment in EMA and EFSA is initiated by the application of a company with a view to release a product onto the Internal Market, the authorisation procedure in ECHA involves an additional stage preceding the risk assessment. In order to identify those substances which ought to be subject to risk assessment, it is necessary to screen the SVHCs and to prioritise respectively.[16] As ECHA has emphasised, the prioritisation process is not risk-based, and the prioritisation stage can therefore not be regarded as risk assessment.[17] Yet as Greim (2010) emphasises, there are different approaches to risk assessment depending on whether existing chemicals or new chemicals are assessed. Whereas the latter are assessed by testing, the former are assessed by deriving available information from a literature search. This is exactly what is done in the prioritisation process of SVHCs, and, more importantly, the prioritisation criteria can be seen as proxies for the criteria used in the risk assessment (such as hazard or exposure).[18]

A first step in the prioritisation is the creation of the Candidate List comprising those substances which possibly become subject to authorisation.[19] In this respect, NCAs, their specific expertise and the work they have done under previous EU substance legislation is crucial, particularly regarding the first substances identified. In order to initiate the authorisation procedure, NCAs submit the so called Annex XV dossier on a specific substance and thereby propose to put this substance on the Candidate List.[20] The proposed SVHCs then need to be agreed upon in the Member State Committee (MSC) of ECHA. It has been pointed out above that there is an essential difference between the EMA and EFSA with regard to the composition of their scientific bodies with representatives of the Member States. As the name suggests, in ECHA's MSC, Member States are directly represented by sending officials from their NCAs in the field of chemicals regulation (Regulation 1907/2006, Article 85). It thus resembles the CHMP in the EMA rather than the GMO Panel in EFSA, which consists of independent experts.[21]

The MSC is the crucial organisational body to decide on the identification of a SVHC, and the Member States are thus able to influence the authorisation process from the beginning. Based on the experience with REACH so far, one may argue that the Member States could try to upload their policy preferences with regard to authorisation.[22] The level of politicisation with regard to SVHC identification seems to be relatively low within the MSC;

with more than 70 substances currently being on the Candidate List, only one SVHC proposed by a NCA has not been agreed upon by the MSC. All in all, it seems that at least the first entries into the Candidate List emerged on a rather ad hoc basis, building to a large extent on the technical work being done prior to REACH.

Consequently, we see differences between the role of the agency in risk assessment in the ECHA case compared to the EMA and EFSA. Because the products to authorise are existing and not newly developed, the role of ECHA differs. This leads to the situation where the risk assessment is preceded by a process of prioritisation, and this in turn affects which actors are actually of influence on the substances that are up for authorisation. Whereas in the case of pharmaceuticals and GMOs, the applicants completely control whether or not their products are up for authorisation (by deciding whether or not to submit their application), in the case of chemical substances the agency – in conjunction with the Member States in the MSC – determines which substances to prioritise, and thus which substances will be assessed for authorisation.

ECHA in Risk Management

The identification of substances combined in the Candidate List is only the first step of the priority setting process. As a second step, substances on the Candidate List will then be gradually included in Annex XIV of the REACH Regulation. Once included in that Annex, the respective substances are 'earmarked' and inevitably become subject to authorisation. In other words, the Annex XIV list denotes those substances that are subject to authorisation and to the process of risk analysis similar to the EMA and EFSA. Although it might well be the case that certain substances remain stuck in the Candidate List – should they never be selected for inclusion in Annex XIV – the mere fact of inclusion in the Candidate List may in itself be regarded as risk management.[23] The manufacturer or importer of a substance on the Candidate List must not only notify the ECHA in case the substance is present in articles (Regulation 1907/2006, Article 7); for substances on the Candidate List, a Safety Data Sheet (SDS) needs to be provided for downstream users (Article 31).[24]

Since the substances subject to authorisation are already on the market, the 'transference' from the Candidate List into Annex XIV brings about a so-called 'sunset date'. This date is the deadline by which the substance must be removed from the market. It can only be produced or manufactured beyond this deadline if an authorisation has been granted. In addition, certain uses of the particular substance can be excluded from the authorisation requirement. Hence, two conditions of Annex XIV inclusion are relevant as instruments of risk management: the setting of the sunset date for a substance, and exemptions in respect of certain uses of this substance. Similar to the EMA and EFSA, the role of ECHA in risk management is formally restricted. As REACH stipulates, ECHA is entitled to recommend to the Commission the substances to be included in Annex XIV, and ECHA shall consult the MSC before issuing such a recommendation (Regulation 1907/2006, Article 58).

To date the ECHA has issued three recommendations, of which two resulted into a Commission Regulation by way of comitology (Commission Regulation 143/2011 and Commission Regulation 125/2012). The first recommendation was substantially altered by the MSC. While the ECHA recommended certain exemptions for four out of the seven substances on the list, the MSC refused to accept these exemptions and the recommendation was adapted respectively.[25] Moreover, this first recommendation also provides evidence that the Commission is not merely rubber-stamping opinions of the agency, as in the case with the EMA. After the MSC demanded changes with regard to exemptions, the Commission deviated from the recommendation with regard to the sunset dates of one substance.[26] In this particularly contentious case concerning the substance HBCDD (Hexabromocyclododecane), the Commission extended the respective sunset date, thereby giving industry more time to prepare the application for authorisation (or to phase out the substance in case no application be filed). Interestingly in this respect, it took about one and a half years to transform the agency recommendation (June 2009) into a decision taken by comitology (February 2011). One important reason was the unresolved conflict between DG Enterprise and Industry and DG Environment as regards the sunset date.[27] While the EMA and EFSA only have one parent DG, due to the ambivalent policy objectives of REACH, two Commission DGs are in charge and this can lead to protracted conflicts within the Commission.

In the actual authorisation process – which follows as soon as companies file applications for the substances included in Annex XIV – another difference compared to the EMA and EFSA arises. In the other two agencies there is a rather clear separation, at least on paper, between risk assessment and risk management. The stage of risk management begins after a respective opinion based on risk assessment 'leaves' the agency and 'enters' the realm of the Commission and comitology. Hence, risk management – as the weighing of policy alternatives – takes place *outside* the organisational boundaries of the agency. For the ECHA, the weighing of alternatives is supposed to take place *within* the organisational boundaries of the agency. Similar to the MSC, two committees are in charge of this: the Committee for Risk Assessment (RAC) and the Committee for Socio-economic Analysis (SEAC).[28] The role of the RAC is to assess risks stemming from the use of a substance including the appropriateness of the risk management measures provided for in the application. The SEAC shall assess socio-economic factors related to possible alternatives for the substance subject to authorisation (such as availability, feasibility and costs). In consequence, the agency opinion sent to the Commission then involves both these elements (Regulation 1907/2006, Article 64).

It remains to be seen how this will all play a role in the actual authorisation of Annex XIV substances. In terms of their composition, both the RAC and the SEAC are in between the CHMP in the EMA and the GMO Panel in EFSA. While the members of the MSC are appointed by the Member States, the members of the RAC and the SEAC are appointed by the Management Board of ECHA. Interview respondents suggest that the RAC and SEAC are true expert committees (in contrast to the MSC) although it is also stated that both committees shall include 'at least one member but not more than two from the nominees of each Member States that has nominated candidates' (Regulation 1907/2006, Article 85). Drawing on the experience of these agencies, one could hypothesise that decision making in comitology procedures will be smooth because the Member States are represented in scientific committees and, accordingly, need not resort to politicising the comitology procedure (as so vividly visible in the GMO authorisation process). However, first evidence of the prioritisation process and the first Annex XIV decision is less clear cut. The substance HBCDD may serve as a clear example here. After the identification of HBCDD as a SVHC in October 2008, the MSC agreed on ECHA's recommendation to further prioritise the substance for Annex XIV inclusion in May 2009. The inclusion was challenged by a number of Member States, in conjunction with industry, who argued that regulatory instruments other than authorisation would be appropriate to deal with HBCDD. However, the dissenting Member States did not vote against the committee decision but instead attached an opinion to the minutes of the MSC meeting to express the dissenting view.[29] As a result, the ECHA recommended including HBCDD into Annex XIV, but the subsequent steps in the prioritisation process proved equally contentious. As indicated above, there was conflict within the Commission with regard to the sunset date which in the end has been pushed back a few months, in contrast to the ECHA recommendation.

Furthermore, the ECHA recommendation provoked conflict among the Member States within the comitology committee and no unanimous decision on the respective Commission regulation was taken. It is interesting to note that whereas these conflicts have not been expressed by voting in the MSC, dissent from the majority opinion is expressed by explicit voting in the comitology committee. Although the division or institutional deadlock that is present in respect of GMO issues is not apparent, the image of comitology as a technical forum in which deliberation prevails over political negotiation also seems inaccurate. In this sense, what can be derived from the early experiences of ECHA is that the authorisation of chemicals does not stand for the development of an epistemic community consisting of technical experts dominating political experts (Eriksson *et al.* 2010). It is rather a politicised process in which technical issues are linked to political interests and national preferences and, as several authors suggest, the agency solution to deal with complex matters of risk is not a panacea per se for effective risk analysis (Chapman 2007; Karlsson 2010).

Conclusion

Through the two centralised authorisation procedures for medicinal products and GMOs, we have outlined that the role of European agencies in risk governance is not limited as risk assessors. Rather, the EMA and EFSA seem to play an important role in both risk assessment and risk management. In the risk assessment phase, the agencies are often presented as the only actors. With regard to the EMA and EFSA, however, risk assessment is shared by the tripartite assessors, namely the agencies, industrial applicants, and national competent authorities. The initial tests of risk assessment are always done at the industry level, and the agencies together with NCAs (although the level of their co-operation varies) finalise risk assessment based on the data and information provided by the industry. While the EMA's extensive interaction with NCAs is viewed as key and in a positive sense, unavoidable in order to adopt an opinion, EFSA involves NCAs only in certain situations for a narrowly defined task — such as environmental risk assessment — during the risk assessment period.[30]

We also see that the ability of EU agencies to influence risk management depends on certain conditions. The Member States' influence on the agency is particularly different compared to their influence on the EMA and EFSA. Since the Member States do have the opportunity to discuss scientific information-related or procedure-related matters in the CHMP in EMA, they feel that they have considerable input into the opinions of the EMA. The Commission formulates draft proposals and they are always adopted through comitology procedures in line with the EMA's opinions. This results in a smooth procedure of the pharmaceuticals authorisation without creating disagreements in the decision-making bodies. By contrast, the comitology committee and the Council have not been able to reach a decision on the Commission's draft proposals on the GMOs authorisation. We argue that this can at least partly be explained by the fact that Member States' interests are not reflected fully at the risk assessment stage.

The fact that EFSA has been able to function as the *de facto* risk manager has held true under the old comitology rules largely because the Commission *had* to adopt a decision after the Council failed to decide. As stated before, it could very well be the case that EFSA's position as the *de facto* risk manager is bound to change with the new comitology rules in which the Commission's obligation to make a final decision becomes optional. Without any institutional reform to handle national disagreements and to accommodate minority views during the risk assessment stage, it may be expected that the comitology procedure remains unable to reach an agreement on the GMOs authorisation. Moreover, since the appeal committee is likely to turn into a 'Council in everything but name' (Hardacre and Kaeding 2011: 15), to some extent it is hard to imagine that the appeal committee will present a solution to overcome the previous inability of the Council to break the standing committee's deadlock. If the Commission does not choose to make a final decision in light of the explicit opposition by a minority group of the Member States, EFSA's opinion does not automatically become the final decision on the GMOs authorisation. Then, it could very well be the case that EFSA's role will be limited to the risk assessment.

When linking this to the ECHA, there seems to be a slight difference because so far the agency has been dealing with substances that have been on the market for a long time. As a result, risk assessment is not as dependent on the applicant's data, and often supported by a wealth of scientific research. However, as already indicated, increasingly the ECHA seems to take up its role in risk assessment by engaging in the identification of SVHC. The actual role of ECHA in the authorisation process still remains to be seen. Nevertheless, when we compare the procedures as we know them with current practices in the EMA and EFSA, we can expect the ECHA to have a considerable impact on risk management because the Member States seem to play a strong role in the agency's scientific committees and the Member States seem largely to agree under comitology procedures. One apparent difference between the EMA and the ECHA is that the involvement of two (somewhat opposing) DGs within the Commission somehow provides a window of opportunity for industry to influence the risk management stage. The very first decision taken under the comitology procedure in this field shows how the heavily lobbied DG Enterprise and Industry managed to win an extended sunset date. This might indicate that the influence of the agency as a risk manager not only depends on the role of the Member States in the authorisation procedure but also on the ties between industry and the Commission.

These three case studies inform us that there is no clear limit to the role of agencies in risk assessment. Moreover, in order for them to play a role in risk management, various national opinions should be heard and harmonized from the beginning of the authorisation procedure. It implies that scientific advice can never be truly separated from political considerations (Levidow *et al.* 2005; Millstone and Van Zwanenberg 2001). Regulatory policy needs expertise but it should be understood that expertise does not always have to take the form of consensus only by scientists. Instead, expertise should incorporate various views and concerns from different domains in the process of risk analysis. A narrow expert consensus may generate expert disagreement, intensifies regulatory conflict and jeopardises the legitimacy of EU decisions (Levidow and Carr 2007).

In advocating the EU agencies as a policy tool to govern risks, these three cases have illustrated that we should not overestimate the possibilities of creating truly independent bodies of experts to make objective risk assessments. The cases illustrate the strong role of industrial applicants and Member State representatives in the risk assessment process. In addition, agencies play a much stronger role in influencing the decisions taken at the risk management stage than anticipated by the wording of the founding regulation. The fact that agency opinions are most likely to be followed up at the decision making stage when Member State opinions have been heard, seems to indicate that the Member States do not readily delegate risk decisions to the European level (be it the Commission, or the agency), and aim to hold at least some control over the risk regulation process.

Notes

1 In contrast to the EU in general, the overall budget and staff numbers of EU agencies increased from EUR 100 million and 1,200 officials in 1998 to EUR 2.4 billion and about 5,000 officials in 2011, see *EUobserver*, 25 October 2010.
2 A seventh committee is to be established in 2012: the Pharmacovigilance Risk Assessment Committee.
3 Interview with a Management Board member.
4 Interviews with two CHMP members.
5 Interviews with two CHMP members.
6 More precisely, the opinion is submitted to the Pharmaceuticals unit within the parent DG Health and Consumers. On 1 March 2010, the responsibility for regulation of medicinal products was transferred from DG Enterprise and Industry to DG Health and Consumers.
7 Interview with EMA staff. See also Gehring and Krapohl (2007); Busuioc (2010).
8 See EFSA website at <http://www.efsa.europa.eu/en/aboutefsa/efsawhat.htm> (accessed on 24 August 2011).
9 Interview with EFSA staff.
10 Conclusion based on the analysis of the minutes of the GMO Panel's plenary meetings between 2004 and 2011. Regarding unanimity, it should be noted that in some draft opinions issues or text are discussed and amended during a plenary meeting. In these cases, adoption of these opinions is deferred to the next plenary meeting.
11 Interview with a GMO Panel member.
12 Interview with a GMO Panel member.
13 This had been the case under the previous comitology rule. Based on Article 291 of the Lisbon Treaty, the new comitology rule (effective since 2011) has replaced the Council to the Appeal Committee in the centralised authorisation procedure. We have no updated empirical information on the situation of referring to the Appeal Committee in case of no opinion from comitology.
14 See *Europeans and Biotechnology in 2005: Patterns and Trends*, Eurobarometer Report 643 (2006).
15 SVHCs are substances which are considered particularly hazardous because, e.g. they cause cancer or accumulate in the environment.
16 Nobody really knows how many SVHCs are actually on the market; estimates range from around 300 to about 1500. See for instance Romano *et al.* (2011) as well as the so called 'SIN list' compiled by environmental NGOs (<http://www.chemsec.org/list> accessed on 26 January 2012). These substances, plus newly discovered SVHCs (such as via the REACH registration procedure), are potential targets for the authorisation.
17 See Minutes of the 14th Meeting of the Member State Committee (MSC-14), MSC/M/014/2010 of 20 October 2010.
18 *Ibidem.*
19 For the current Candidate List, see <http://echa.europa.eu/web/guest/candidate-list-table> (last accessed on 2 February 2012).
20 The Annex XV is thus the main document in the prioritisation process since it can be regarded as risk assessment as mentioned by Greim (2010). By referring to respective studies it provides for the justification of why the substance is deemed hazardous and should be subject to authorisation.

21 The members of the MSC are well aware that they represent Member States' interests in the MSC even though they can be considered as policy experts (see Minutes of the 15th Meeting of the Member State Committee (MSC-15), MSC/M/05/2010 of 1 February 2011).

22 As indicated by interview sources, the substances dealt with in Annex XV dossiers reflect experiences and thus policy preferences of Member States' NCAs. Sweden for instance, which is generally a proactive Member State in chemicals regulation, has a particular focus on so-called brominated flame retardants (see Eriksson *et al.* 2010). Consequently Sweden has acted as Rapporteur for such substances before and after REACH. In a similar vein, Germany is particularly concerned with PBT substances and has by far submitted the most Annex XV dossiers for these substances. See also Heyvaert (2010: 227) on this point.

23 The reason why not all substances on the Candidate List are included into Annex XIV has to do with the administrative capacities of ECHA. Since every inclusion yields applications for authorisation coming from the industry, the agency's capacity to handle these applications needs to be taken into account (Regulation 1907/2006, Art. 58(3)).

24 It has been stressed by interview respondents that this is an important reason to include substances into the Candidate List. The Candidate List might then serve the purpose of generating information even though the substance may not be prioritised for Annex XIV inclusion.

25 See Recommendation of the European Chemicals Agency (ECHA) of 1 June 2009 for the inclusion of substances in Annex XIV (the list of substances subject to authorisation) of Regulation 1907/2006.

26 In addition, the Commission removed one substance from the list altogether. The Commission stated that after the prioritisation stage of the ECHA authorisation procedure, the substance has been included into the international Protocol on Persistent Organic Pollutants. This inclusion triggered the respective obligations for the EU related to Regulation 850/2004 of the European Parliament and of the Council of 29 April 2004 on persistent organic pollutants and amending Directive 79/117, OJ 2004 L 158/7 (see Commission Regulation 143/2011).

27 As indicated above, DG Enterprise and Industry succeeded in negotiating an extended sunset date (Interview with a representative of the Dutch authorities).

28 It remains to be seen how the naturally close relationship between the RAC the NCAs evolves in the practice of dealing with authorisation applications (see Minutes of the 1st Meeting of Committee for Risk Assessment (RAC-1) 29–31 January 2008, RAC/M/01/2008 Final).

29 Minutes of the 8th Meeting of the Member State Committee (MSC-8) 18–20 May 2009, MSC/M08/2009 Final. Interesting in this respect is the remark made by the chair of the MSC whereas 'the MSC has to look at sound scientific arguments and not political arguments and consider the arguments that are relevant for this specific step in the authorisation process' (MSC/M08/2009 Final, p. 9).

30 Interview with a CHMP member of the EMA.

Bibliography

Borrás, S. (2006) 'Legitimate Governance of Risk at the EU Level? The Case of Genetically Modified Organisms', *Technological Forecasting & Social Change*, 73: 61–75.

Braun, D. and Gilardi, F. (eds) (2006) *Delegation in Contemporary Democracies*, New York: Routledge.

Busuioc, M. (2010) *The Accountability of European Agencies: Legal Provisions and Ongoing Practices*, Delft: Eburon.

Chalmers, D. (2005) 'Risk, Anxiety and the European Mediation of the Politics of Life', *European Law Review*, 30(5): 649–74.

Chapman, A. (2007) *Democratizing Technology. Risk, Responsibility and the Regulation of Chemicals*, London/Sterling: Earthscan.

Commission of the European Communities (2002) *The Operating Framework for the European Regulatory Agencies*, COM (2002) 718 final, 11 December, Brussels.

Eriksson, J., Karlsson, M. and Reuter, M. (2010) 'Technocracy, Politicization, and Noninvolvement: Politics of Expertise in the European Regulation of Chemicals', *Review of Policy Research*, 27(2): 167–85.

Ernst & Young et Associés (2010) *European Commission Evaluation of the European Medicines Agency – Final report*, Evaluation commissioned by and presented to the DG Enterprise and Industry.

European Medicines Agency (2008) *The Linguistic Review Process of Product Information in the Centralised Procedure – Human*, Doc.Ref: EMEA/5542/02/Rev 3.3, 15 October, London.

Foth, H. and Hayes, A.W. (2008) 'Background of REACH in EU Regulations on Evaluation of Chemicals', *Human & Environmental Toxicology*, 27: 443–61.

Gehring, T. and Krapohl, S. (2007) 'Supranational Regulatory Agencies between Independence and Control: The EMEA and the Authorization of Pharmaceuticals in the European Single Market', *Journal of European Public Policy*, 14(2): 208–26.

Greim, H. (2010) 'Chemical Risk Assessment in Toxicological Perspective', in J. Eriksson, M. Gilek and C. Rudén (eds), *Regulating Chemical Risks. European and Global Challenges*, Dordrecht et al.: Springer, 121–31.

Hardacre, A. and Kaeding, M. (2011) 'Delegated & Implementing Acts: The New Comitology', *EIPA Essential Guide*, 3, Maastricht: European Institute of Public Administration.

Heyvaert, V. (2010) 'Regulating Chemical Risk: REACH in a Global Governance Perspective', in J. Eriksson, M. Gilek and C. Rudén (eds), *Regulating Chemical Risks. European and Global Challenges*, Dordrecht et al.: Springer, 217–37.

Hood, C., Rothstein, H. and Baldwin, R. (2002) *The Government of Risk: Understanding Risk Regulation Regimes*, New York: Oxford University Press.

Hutter, B. and Power, M. (eds) (2005) *Organizational Encounters with Risk*, Cambridge: Cambridge University Press.

Jasanoff, S. (2005) *Design on Nature. Science and Democracy in Europe and the United-States*, Princeton: Princeton University Press.

Karlsson, M. (2010) 'The Precautionary Principle in EU and US Chemicals Policy: A Comparison of Industrial Chemicals Regulation', in J. Eriksson, M. Gilek and C. Rudén (eds), *Regulating Chemical Risks. European and Global Challenges*, Dordrecht et al.: Springer, 239–65.

Krapohl, S. (2004) 'Credible Commitment in Non-independent Regulatory Agencies: A Comparative Analysis of the European Agencies for Foodstuffs and Pharmaceuticals', *European Law Journal*, 10(5): 518–38.

Levidow, L. and Carr, S. (2007) 'Europeanising Advisory Expertise: The Role of "Independent, Objective and Transparent" Scientific Advice in Agri-biotech Regulation', *Environment and Planning C: Government and Policy*, 26(6): 880–95.

Levidow, L., Carr, S. and Wield, D. (2005) 'European Union Regulation of Agri-biotechnology: Precautionary Links between Science, Expertise and Policy', *Science and Public Policy*, 32(4): 261–76.

Lindgren, K.O. and Persson, T. (2008) 'The Structure of Conflict over EU Chemicals Policy', *European Union Politics*, 9(1): 31–58.

Löfstedt, R. (2005) *Risk Management in Post-Trust Societies*, London: Earthscan.

Majone, G. (2002) 'Delegation of Regulatory Powers in a Mixed Polity', *European Law Journal*, 8(3): 262–75.

——(2010), 'Strategic Issues in Risk Regulation and Risk Management', in OECD (ed.), *Risk and Regulatory Policy: Improving the governance of risk*, Paris: OECD, 93–131.

Millstone, E. and Van Zwanenberg, P. (2001) 'Politics of Expert Advice: Lessons from the Early History of the BSE Saga', *Science and Public Policy*, 28(2): 99–112.

Pesendorfer, D. (2006) 'EU Environmental Policy under Pressure: Chemicals Policy Change between Antagonistic Goals?', *Environmental Politics*, 15(1): 95–114.

Romano, D., Santos, T. and Gadea, R. (2011) 'Trade Union Priority List for the REACH Authorization', *Journal of Epidemiology and Community Health*, 65: 8–13.

Sabel, C. and Zeitlin, J. (2008) 'Learning from Differences: The New Architecture of Experimentalist Governance in the EU', *European Law Journal*, 14(3): 271–327.

Selin, H. (2007) 'Coalition Politics and Chemicals Management in a Regulatory Ambitious Europe', *Global Environmental Politics*, 7(3): 63–93.

Skogstad, G. (2003) 'Legitimacy and/or Policy Effectiveness: Network Governance and GMO Regulation in the European Union', *Journal of European Public Policy*, 10(3): 321–38.

Van Asselt, M.B.A. and Vos, E. (2006) 'The Precautionary Principle and the Uncertainty Paradox', *Journal of Risk Research*, 9(4): 313–36.

Van Asselt, M.B.A. and E. Vos (2008) 'Wrestling with Uncertain Risks: EU Regulation of GMOs and the Uncertainty Paradox', *Journal of Risk Research*, 11(1-2): 281–300.

Van Asselt, M.B.A., Vos, E. and Rooijackers, B. (2009) 'Science, Knowledge and Uncertainty in EU Risk Regulation', in M. Everson and E. Vos (eds), *Uncertain Risks Regulated*, Abingdon: Routledge-Cavendish, 359–88.

Vos, E. (2000) 'EU Food Safety Regulation in the Aftermath of the BSE Crisis', *Journal of Consumer Policy*, 23(3): 227–55.

——(2003) 'Agencies and the European Union', in T. Zwart and L. Verhey (eds), *Agencies in European and Comparative Law*, Oxford: Intersentia, 113–54.

Vos, E. and Wendler, F.A. (eds) (2006). *Food Safety Regulation in Europe. A Comparative Institutional Analysis*, Antwerp/Oxford: Intersentia.

Wynne, B. (2001) 'Creating Public Alienation: Expert Cultures of Risk and Ethics', *Science as Culture*, 10(4): 445–81.

9 The Precautionary Principle in Court

An Analysis of Post-Pfizer Case Law

Anne-May Janssen and Marjolein van Asselt

Introduction

When the European Courts are presented with a risk regulation case that involves scientific uncertainty, they face the difficult task of judging whether or not it is legitimate to apply the precautionary principle. Although there are several descriptions and definitions of the precautionary principle, it lacks a single, clear and universally accepted definition (Percival 2005). Article 15 of the Rio Declaration is the most commonly known statement, which holds that '[w]here there are threats of serious or irreversible damage, lack of full scientific certainty shall not be used as a reason for postponing cost-effective measures to prevent environmental degradation' (UN Economic and Social Council 1997). It is designed 'to address the existence of scientific uncertainty in areas where our failure to anticipate future harm may lead to disaster' (UN Economic and Social Council 1997; Kaiser 1997: 202). Thus, it is possible that precautionary action must be taken, even if fears might prove to be unwarranted (Kaiser 1997). Within the European Union, the precautionary principle has become a general principle of EU law, as the European Commission acknowledged (European Commission 2002a) and is established by case law.[1]

It has been argued that the precautionary principle cannot be applied to every type of risk (Van Asselt and Vos 2006, 2008). The principle covers significant and serious risks that could cause irrevocable damage. Its application is only logical when risk and uncertainty interact (De Sadeleer 2007). Thus, it is intended to address so-called 'uncertain risks' (Van Asselt and Vos 2006, 2008):[2] situations in which there are serious reasons to believe that there may be danger, but the scientific evidence is neither sufficient to substantiate that danger nor to refute suspicions of that danger arising. When attempting to predict the consequences and impact of new technological developments on society, 'even experts grope in the dark' (WRR 2009: 103). Uncertain risk problems emerge when 'activities are undertaken with consequences that can be anticipated only partially, if at all' (WRR 2009: 102). Thus, '[t]he notion of uncertain risks refers to possible, new, imaginable hazards, with which society has no or limited experience' (Van Asselt and Vos 2008: 281).

Also, 'it is generally agreed that uncertainty is the essence of the precautionary principle' (Van Asselt and Vos 2006: 314). In order to trigger the application of the precautionary principle therefore, the uncertainty condition must be met. This means that scientific uncertainty regarding the risk must exist for the precautionary principle to be applicable.

The lack of a clear and accepted definition of the precautionary principle has led to extensive debate, if not controversy. Critics have argued that the lack of a clear definition inhibits any sound application of the principle. In addition, its vagueness arguably makes the principle incoherent and arbitrary (Percival 2005; Sandin 2006a). Moreover, it is not specified how significant the harm must be nor what is constituted by 'cost-effective measures' (Percival 2005; Sandin 2006a). Article 7(2) of Regulation 178/2002 states that any measure taken in the spirit of the precautionary principle must be proportionate, no more restrictive to trade than necessary and should be reviewed in reasonable time (European Commission 2002b). However, it is not specified as to what constitute proportionate or sufficiently restrictive measures. In its *Communication on the precautionary principle*, the Commission states that the EU 'has the right to establish the level of protection [...] that it deems appropriate' (European Commission 2002a: 3). Consequently, critics such as Hahn and Sunstein have argued that the precautionary principle provides no direction at all (Hahn and Sunstein 2005).

The leverage the European Courts have in applying the principle has not gone unnoticed. In their article 'The Precautionary Principle and the Uncertainty Paradox' van Asselt and Vos posed the question whether the ruling of the European Court of First Instance (CFI) in Pfizer Animal Health SA *v*. Council of the European Union (hereafter 'the Pfizer Case') could lead to an erosion of the precautionary principle (compare Forrester and Hanekamp 2006 and Rogers 2011, who also analysed the use of the precautionary principle in this case). Van Asselt and Vos argued that the CFI constructed its own definition of uncertainty. In this way, the Court ruled in favour of application of the principle. In other words, the Court leveraged the interpretative and defining openness, which opens the door for a wide array of problems (2006). In the Pfizer Case, the Court interpreted uncertainty as contrasting scientific opinions to justify the application of the precautionary principle (ibid.). Reading it as a precedent, the precautionary principle could be applied whenever one qualified scientist takes a diverging opinion. Since such a dissenting opinion may be found in nearly all uncertain risk cases, van Asselt and Vos feared that such an interpretation and application of uncertainty would empty the precautionary principle. This is the question addressed in this chapter.

We first summarise the Pfizer Case in more detail and particularly the CFI's ruling. Then, three post-Pfizer cases are discussed, including the events that led to these three cases being brought before the Court. The patterns and inconsistencies in dealing with uncertainty are analysed here and it is argued that these inconsistencies complicate the application of the precautionary principle. The most prominent inconsistencies indeed suggest a *Pfizer* precedent. In addition, other inconsistencies are apparent: lack of risk assessment, a disregard for the temporary nature of precautionary measures and problematic, analogical handclapping. There is a gap between the procedures of risk assessment and the review of risk assessments. Finally, there are inconsistencies in the way in which the Court reviews similar cases. To conclude, a few options open to the Court to improve the current problematic approach to the precautionary principle are discussed.

Pfizer: A Precedent?

The Pfizer Case has been important in the history of the precautionary principle. It was in this case that the CFI discussed at length the interpretation and the correct application of the precautionary principle for the first time (Case Law Analysis 2003; Da Cruz Vilaça 2004). On 17 December 1998 Council Regulation 2821/98 was adopted, withdrawing the market authorisation of certain antibiotics used as additives for feedstuffs. These antibiotics served as growth promoters for animals. Pfizer Animal Health (hereafter 'Pfizer') is a pharmaceutical company producing and marketing virginiamycin, an antibiotic that was banned by Regulation 2821/98. At the time of its adoption, Pfizer was the only virginiamycin producer in the world.[3] Pfizer challenged the regulation before the CFI and sought annulment of that regulation.

Antibiotics in Feed as an Uncertain Risk

An antibiotic is a substance that acts at the level of the metabolism of bacteria. They are used to treat bacterial infections, both in humans as well as in animals. Resistance of bacteria to antibiotics, i.e. the increased capability to resist the therapeutic effects, can present a serious risk for public health, since these bacteria can no longer be treated with the respective antibiotic. For many years, antibiotics have served as growth promoters and were added in low doses to animal feedstuffs. This practice causes animals to grow faster and gain more weight. As a result, they need less time and less food to reach the desired weight for slaughter.[4]

It has been indisputably demonstrated that using antibiotics as growth promoters tends to encourage the development of resistant bacteria in animals. It is very difficult, however, to assess whether or not there is a risk to human health as a result of transferring antibiotic resistance from animals to humans. It is assumed that antibiotic resistance is transferrable from animals to humans. At the time of the debate on antibiotics used as growth promoter no human health problems had been attributed to antibiotic resistant bacteria as a result of using growth promoters (Case Law Analysis 2003; Forrester and Hanekamp 2006). 'However, no producer could ever prove this would never happen, or could never happen', (Forrester and Hanekamp

2006: 304). At that point in time, the scientific community largely recognised the existence of a potential link between the practice of using antibiotics as growth promoters in animals and the development of resistance to those antibiotics in humans.[5]

Regulation 2821/98

In January 1998, Denmark notified the Commission of its decision to ban the use of the antibiotic virginiamycin as growth promoter. This decision was based on a report from the Danish National Veterinary Laboratory, which considered it probable that virginiamycin could be transferred from animals to humans. The Commission submitted the Danish report to the Scientific Committee on Animal Nutrition (SCAN) (Van Asselt and Vos 2006). SCAN was asked by the Commission to give an opinion 'on the question "whether or not" the use of the virginiamycin as a growth pro-moter constitutes a public health risk or could constitute such a risk in the future' (Van Asselt and Vos 2006: 320; SCAN 1998). The conclusion of SCAN was that 'the use of virginiamycin as a growth promote does not con-stitute an immediate risk to public health' (1998). Making reference to the pre-cautionary principle, the Commission deviated from this expert opinion (Van Calster and Lee 2003) and proposed a ban of the antibiotic. The Eur-opean Council agreed with the Commission (Van Asselt and Vos 2006) and, consequently, on 17 December 1998 it adopted Regulation 2821/98 prohi-biting the use of four antibiotics in animal feedstuffs: bacitracin zinc, tylosin phosphate, spiramicin and virginiamycin. At the time of adoption, there was no scientific proof to establish a link between any of the four mentioned antibiotics and development of resistance in humans to these antibiotics.[6]

The Ruling of the Court

The risk to be assessed in the Pfizer Case was the possibility that adding virgi-niamycin to animal feedstuffs would lead to adverse effects on human health in terms of antibiotic resistance. If there was a potential risk of transferring anti-biotic resistance[7] from animals to humans, this could reduce the effectiveness of certain medical products designed for humans (Da Cruz Vilaça 2004).

In their request for annulment, Pfizer's grounds related to incorrect appli-cation of the precautionary principle and errors made in the risk assessment and management. This forced the Court to investigate the scientific arguments made by both parties (Van Asselt and Vos 2006). The Court eventually upheld the ban in its ruling. It is the argumentation for the application of the precautionary principle that has given rise to much debate.

Van Asselt and Vos argued that the Court was in a deadlock, struggling with its role in the scientific debate. The Court stated that '[i]t is not for the Court to assess the merits of either of the scientific points of view argued before it'.[8] It continued, however, to state that:

the Court nevertheless finds that the parties' arguments, supported in each case by the opinions of eminent scientists, show that there was *great uncertainty*, at the time of the adoption of the contested regulation, about the link between the use of virginiamycin (...) and the development of (...) resistance in humans (emphasis added).[9]

The Court then highlighted that 'at the time when the contested regulation was adopted, other scientists and specialist bodies had taken a different view from that of SCAN and the experts called by Pfizer'.[10] However, at the same time the Court stated that there was:

> *sufficient reliable and cogent scientific evidence* (...) to conclude that there was *a proper scientific basis for a possible link* between the use of virginia-mycin (...) and the development of (...) resistance in humans (emphasis added).[11]

Moreover, the Court concluded that 'the Community institutions had a scientific basis on which to reach a decision, since they could draw on some results of the most recent scientific research on the matter'.[12]

Van Asselt and Vos argued that the interaction pattern between science, policy makers and judges could be characterised as the uncertainty paradox: 'while on the one hand *great uncertainty* is emphasised, at the same time, it is suggested that *sufficient reliable and cogent scientific evidence* and *a proper scientific basis* were available' (Van Asselt and Vos 2006: 326). Thus, the Court resorted to experts to draw final conclusions about the existence of the risk. At the same time, the Court interpreted uncertainty as contrasting scientific opinions to justify the application of the precautionary principle, on which basis it upheld the ban of virginiamycin. Thus, the Court equated scientific uncertainty – the core element of the precautionary principle – with diverging opinions between experts (Van Asselt and Vos 2006). This view is also held by Forrester and Hanekamp, who argued that '[t]he Court based its decision not on an examination of the scientific merits, but on the fact that a con-troversy existed between scientists' (2006: 307). Reading it as a precedent, the precautionary principle might be applied whenever one qualified scientist holds a diverging opinion. Since these can be found in nearly all uncertain risk cases, van Asselt and Vos, and Forrester and Hanekamp argue, albeit in different veins, that such an interpretation and application of uncertainty renders the precautionary principle an empty one.

Post-Pfizer Case Law on the Precautionary Principle

It has largely been up to the Courts to clarify the appropriate application of the precautionary principle. Due to its litigious character, the interpretation of the principle has been the single most determining aspect of its jurisprudence here. Stokes (2008) argued that the Courts have not always been consistent

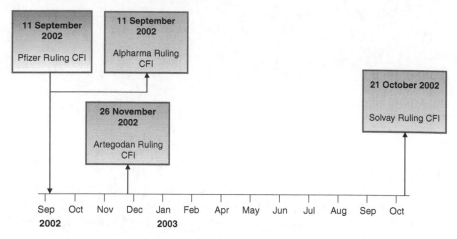

Figure 9.1 Timeline Post-Pfizer case law

in defining the relationship between uncertainty and the precautionary principle: shifts in the judicial interpretation of precaution can be observed (Stokes 2008). In the Pfizer Case, the Court constructed its own definition of uncertainty to warrant the application of the precautionary principle. We sought to investigate whether or not the ruling in the Pfizer Case had influenced subsequent case law in ways that are problematic. To this end, three cases brought before the CFI soon after the Pfizer Case and concerning the application of the precautionary principle (see Figure 9.1) have been researched for the purposes of establishing whether and how the Pfizer Case served as a precedent in those cases. Should this be the case, follow-up research into more recent cases where the precautionary principle has been applied could be investigated to examine whether problematic patterns are still visible.

Several patterns and inconsistencies in the interpretation and application of the precautionary principle are identified. The cases are introduced first, before the patterns and inconsistencies are described.

Alpharma Inc. *versus* Council of the European Union

Alpharma produced and marketed bacitracin zinc, which is also an antibiotic used as an additive in animal feedstuffs. Bacitracin zinc was authorised under Directive 70/524. On 2 February 1998, Sweden submitted a request to amend that Directive, seeking to withdraw the market authorisation of bacitracin zinc. Attached to the request was a report that addressed the risk of using bacitracin zinc in both human and animal treatment. Though indicating that people who use bacitracin zinc on a daily basis may be sensitised, the report acknowledged that there is not enough information to assess the possible risk of the antibiotic to human and animal health. This report was discussed at a SCAN meeting but no SCAN opinion was requested by the Commission.[13]

On 6 June 1998, Alpharma applied for renewal of market authorisation of bacitracin zinc. On 5 November 1998, Alpharma was informed that the Commission had drawn up a draft proposal that would add bacitracin zinc to the list of banned additives for feedstuffs. On 17 December 1998, the Council adopted the contested Regulation 2821/98, thereby banning bacitracin zinc.[14] In response, Alpharma brought proceedings before the CFI seeking annulment of that regulation. Similar to the Pfizer Case, the question before the Court was whether the use of bacitracin zinc as an additive to animal feed posed a potential risk to human health. Due to the questions the Court was asked, it had to deal with issues of scientific uncertainty. In September of 2002, the Court ruled in favour of the Commission, and the Council upheld the ban.

Solvay Pharmaceuticals BV *versus* Council of the European Union

Nifursol is an additive used in feedstuffs and belonging to the group of nitrofurans, a class of antibiotic drugs. It is used to prevent the occurrence of a parasitic disease in turkeys.[15] Nifursol was authorised in 1988. The 1995 Regulation No. 2377/90, which decreed that it was no longer permitted to administer nitrofurans as veterinary medicinal products to food-producing animals, did not include Nifursol. The ban was based on an examination by the Committee for Veterinary Medicinal Products of four specific nitrofurans substances: nitrofurazone, nitrofurantoine, furaltadone and furazolidone. In 1999, Nifursol was subject of new authorisation under Article 9h of Directive 70/524, for which Solvay applied. The Commission pointed out that Nifursol belongs to the nitrofurans group and that there needs to be consistency in the rules for the substances belonging to that group. The provisional authorisation for Nifursol was replaced on 16 November 1999, with an authorisation linked to the manufacturer, Solvay Pharmaceuticals, through Regulation 2430/1999. The UK Veterinary Medicines Directorate (VMD) forwarded to Solvay an expert opinion on 22 May 2000.[16] The VMD acknowledged that the available data concerning Nifursol are incomplete. However, it cannot rule out that it could pose a risk. On 11 October 2001, SCAN adopted an opinion on Nifursol. It concluded that there was not enough data available but that the safety of Nifursol could not be guaranteed. On 23 September 2002, the Council adopted Regulation 1756/2002 whereby the authorisation of the additive was withdrawn.[17] In response to that regulation Solvay Pharmaceuticals brought proceedings before the CFI. The Court ruled in favour of the EU institutions, upholding the withdrawal of the authorisation.[18]

Artegodan GMBH and Others *versus* Commission of the European Communities

The applicants of this case were holders of market authorisation for medicinal products that contain amphetamine-like anorectic agents. These substances influence the body at the central nervous system level to speed up the feeling of satiation. These substances were used in the treatment of obesity for many years in a number of Member States.[19] The authorisations were originally issued by the competent national authorities. On 17 May 1995, the Federal Republic of Germany made a referral to the Committee for Proprietary Medicinal Products (CPMP) of the European Agency for the Evaluation of Medicinal Products under Article 12 of Directive 75/319. Germany expressed its concerns that certain centrally-acting anorectics were suspected of inducing primary pulmonary hypertension (PPH). PPH occurs when the blood pressure is abnormally high in the arteries of the lungs and the right side of the heart needs to work harder than normal. Over time, this causes the heart muscles to weaken and they may eventually fail (MedlinePlus 2011). On 17 July 1996 the CPMP issued its final opinion on the anorectics agents mentioned above and recommended maintaining market authorisations under the condition that a few amendments were made in the description of the product characteristics.[20] The Commission instructed the Member States concerned through the decision of 9 December 1996 that, subject to amendments in the product summaries, the marketing authorisations were granted.

The Belgian Ministry for Social Affairs, Public Health and the Environment expressed its concerns about the risk of cardiac valve disorders. In response to these concerns, the CPMP initiated the procedure under Article 13 of Directive 75/319 with respect to amfepramone. There was also an evaluation of phentermine conducted. In its final opinion of 31 August 1999, the CPMP recommended withdrawing the market authorisations of medicinal products containing amfepramone and phentermine.[21] Concerning the substances clobenzorex, fenproporex and norpseudoephedrine, the CPMP – based on the same reasoning as with amfepramone and phentermine – recommended withdrawal of market authorisation.[22]

On 9 March 2000, the Commission adopted three decisions ordering the Member States to withdraw the national marketing authorisations for several medicinal products containing amphetamine-like anorectic agents: Decision C(2000) 453 concerned the withdrawal of amfepramone; Decision C(2000) 452 concerned phentermine; and Decision C(2000) 608 withdrew authorisation for clobenzorex, fenproporex and norpseudoephedrine. The applicants brought proceedings before the Court, which ruled in favour of the applicants and ordered the Commission to annul all three decisions.[23]

Patterns and Inconsistencies

Analysis of the cases identifies several patterns and inconsistencies. Certain definitions and interpretations of uncertainty are problematic. The examined case law reveals not only that multiple definitions of what constitutes uncertainty may be observed (inconsistency), but also that the various ways in which uncertainty is dealt with are problematic. In addition, there are multiple definitions of the relationship between uncertainty and the precautionary principle.

Dealing With Uncertainty

In situations of uncertainty, the precautionary principle may be invoked. In other words, in order to apply the precautionary principle, uncertainty must be established. Consequently, how uncertainty is defined is of significant importance. In practice, the Court has been developing this definition. The case law examined indicates that the Court is struggling with this task and this has led to two problematic patterns in constructing uncertainty.

Constructing Uncertainty Through Contrasting Scientific Opinions

The Pfizer Case established a definition of uncertainty based on contrasting scientific opinions. In its ruling, the Court very clearly referred to the fact that the opinions of experts and scientists were divided as to whether or not there is a link between virginiamycin usage in animal feed and the development of antibiotic resistance in humans.[24] The Court then construed uncertainty as diverging opinions between experts in order to legitimise the application of the precautionary principle. A similar pattern can be established when reviewing the Alpharma case (Van Asselt and Vos 2006, 2008). Both Alpharma and the Council concurred that more research into a possible link between the use of bacitracin zinc and the transfer of antibiotic resistance to humans was necessary. The point on which they disagreed was whether, and if so to what extent, antibiotic resistance could be transferred to humans through the use of bacitracin zinc.[25] The Court acknowledged that '[t]he possibility and the probability of such transfer and the risk it may entail for public health continue to give rise to *argument in scientific circles*'[26] (emphasis added). The Court asserted that:

> [i]t is common ground between the parties [...] that at the time when the measure was adopted, the transfer and development of such resistance had not yet been scientifically established in respect of bacitracin zinc.[27]

The Court initially attempted to keep the argument of contrasting scientific opinions out of its ruling by stating that '[i]t does not matter that [...] there are differences of opinion between scientists'.[28] However, though the Court tried to avoid using diverging scientific views in its ruling, the argument still finds its way into its justification for the precautionary principle. The Court recognised that the withdrawal of the authorisation to market bacitracin zinc 'was established on extremely complex scientific and technical assessments over which scientists have widely diverging views'.[29] On the question of whether bacteria are naturally resistant to bacitracin zinc, the Court held that:

> it is clear from the documents before the Court that, at the time when the contested regulation was adopted, *scientific opinion was sharply divided*

on the question of natural resistance in the particular case of bacitracin[30] (emphasis added).

So in the end, the Court used scientific dissensus to construct uncertainty about the risk, which was then used to justify the application of the precautionary principle. Analysis of both *Pfizer* and *Alpharma* suggests that establishing uncertainty with reference to contrasting scientific opinions is a recurrent pattern.

Constructing Risk by Analogical Handclapping

Uncertainty in the Alpharma Case was not only constructed through scientific dissensus. The Court also argued in terms of analogy with other antibiotics. In the Alpharma case there was no SCAN opinion present on bacitracin zinc. Instead, the Court referred on several occasions to the SCAN opinion on virginiamycin; that which was used in the Pfizer Case. Since there were no specific assessments of the effects of bacitracin zinc the Court held that:

> it was possible, on the basis of the available scientific knowledge relating to those other antibiotics, to accept that the transfer mechanisms were *similar* for all antibiotics and that the transfer of resistance to bacitracin zinc was therefore highly probable[31] (emphasis added).

In Alpharma, the Court used SCAN's scientific opinion on virginiamycin and argued that they supported the conclusion of the EU institutions of 'a risk associated with those products'.[32,33] Thus, the Court relied on previous analyses of other antibiotics to conclude that there was a possible risk in the case of bacitracin zinc. By doing so, the Court concluded that there was actual risk and so, did away with any uncertainty.

Analogy was also applied in the Solvay Case on Nifursol. It is important to note that the VMD had not examined Nifursol. SCAN emphasised that 'the veterinary medicinal products committee had not examined Nifursol in its opinions issued between 1990 and 1995, which resulted in the ban in 1995 on the use of all nitrofurans as veterinary medicinal products' (SCAN 2001).[34] Notwithstanding the lack of a risk assessment on Nifursol, on 22 May 2000 the VMD forwarded an expert opinion to Solvay. In that opinion it was argued that 'it is proven that certain nitrofurans are genotoxic and that that risk is thought to be associated with the presence in the molecule of a "group 5-nitro"'.[35] On this basis, the VMD held that 'since Nifursol has the same molecular structure, it is [...] also likely to present a risk of genotoxicity'.[36] On 11 October 2000, SCAN issued its opinion on Nifursol in which it argued that the substances that were examined all demonstrated specific characteristics and that there were differences between the substances. The molecular structures of the substances were 'very different from each other and therefore had different effects'.[37] In addition, 'it has never been proven that the presence of a "group 5-nitro" in the molecular structure, characteristic of nitrofurans, is the cause of such risks'.[38] The Court, in its turn argued,

> it is *not disputed by the Council and the Commission that the mere fact that Nifursol belongs to the nitrofurans group was not sufficient*, in the absence of a specific assessment of the safety of that substance, to conclude that it posed the same risks as those cited by the veterinary medicinal products committee in respect of furazolidone and nitrofurazone[39] (emphasis added).

However, the Court then asserted that:

> it should also be noted that the VMD stressed [...] that it may be supported that the risk of genotoxicity, established in the case of certain nitrofurans, is *associated with the presence of a group 5-nitro* in the molecular structure of those substances. The VMD inferred from this that *Nifursol was also suspected of posing such a risk*[40] (emphasis added).

The Court further stated that in the general domain of additives for feedstuffs 'the existence of *solid evidence* which, while not resolving the scientific uncertainty, may reasonably raise doubts as to the safety of a substance'[41] (emphasis added). On this basis the Court ruled in favour of the Council and the Commission.

Also in *Solvay*, therefore, the Court used analogy as a base for its ruling. It claimed that all antibiotics and all nitrofurans have similar characteristics and should be treated in the same way. This is problematic for various reasons. First, it suggests that when uncertainty has been established for one specific substance, there is a general claim of uncertainty. Second, when in previous cases uncertainty has been established to uphold precautionary bans, analogical reasoning opens the door for more bans. This endangers future authorisation procedures for feed additives. The precautionary principle could become a tool to prohibit marketing of all feed additives as soon as uncertainty has been established, and then used as a basis to justify a precautionary ban. When substance-specific characteristics are no longer needed in the risk assessment and commonalities suffice, the Court then ignores the rules that it had laid down previously for such assessments. Jasanoff's (2005) view on process and product framing[42] is useful when discussing these findings. In the process framing, the risks lay in the techniques and thus pertain to the class of products associated with those techniques. In the product framing, the risks lay in its specific application, and are thus associated with a particular product. In this case, we see a product-based regulatory approach clashing with a process-based line of reasoning (risks for a class of products). Jasanoff argues that the political culture in Europe is more inclined towards process framing. From this perspective, the inconsistency in *Solvay* can be understood partly as the regulators' struggles with a product-based approach.

Concluding Remarks

Analysis of the *Pfizer* and post-*Pfizer* case law demonstrates the existence of two patterns in the Courts' attempt to give form and substance to 'uncertainty': constructing uncertainty on the basis of scientific dissensus, and constructing risk by analogical handclapping. The first pattern is in accordance with the precedent set in the *Pfizer* Case. In *Pfizer* and *Alpharma* the Court used scientific dispute to establish uncertainty as to the risk involved. In such a way and as a precedent, the precautionary principle may be applied whenever there is one qualified dissenting opinion.

The second pattern demonstrates that analogy between substances is used to establish that there is a risk, thereby discarding uncertainty. Tetlock's warning (2005) about analogical handclapping addressed to experts, also seems to be very relevant in the context of the European Court: 'overfitting the most superficially applicable analogy [...] is a common source of error' (as cited in Van Asselt *et al.* 2010: 116). Furthermore, a product-based regulatory regime is at odds with this kind of analogical reasoning, which results in class-based[43] risk management. Without reform of the regulatory regime, analogical reasoning is unacceptable.

How Much Uncertainty Allows For Precautionary Measures?

The Court must determine whether a risk assessment has been carried out and whether it is sufficient because procedural requirements demand that a risk assessment precedes risk management. Scientific evaluation precludes that the principle be used as a pretext for other, unwarranted measures (Alemanno 2007). Should no scientific evaluation be performed, it becomes more difficult to prevent. At the same time, in situations of uncertainty experts cannot provide certainty about the risk involved and therefore, a classic risk assessment as a quantitative calculation informed by probability and effect estimates is anyway not possible. Consequently, it is not straightforward to satisfy the risk assessment obligation in situations of uncertainty.

Secondly, there is no recognised risk assessment method in the EU legal order (De Sadeleer 2007). The Court has established that a risk management measure must be 'adequately backed up by the scientific data available at the time the measure was taken'.[44] However, the Court does not explain what 'adequately backed up' means in practice. In case law the Court has determined that application of the precautionary principle 'requires a scientific assessment of risks as exhaustive as possible on the basis of scientific opinions based on principles *of excellence, transparency and independence*'[45] (emphasis added) and that precautionary measures be 'based on as complete a risk assessment as possible'.[46] Thus, the European Courts have confirmed that precautionary measures require a risk assessment (Stokes 2008). Additionally, it is also clear that expert opinions by bodies such as SCAN, VMD and the CPMP are accepted as risk assessment, even where the opinion finds that there to be too much uncertainty to do a risk assessment. The fact that experts have been asked to consider the risks seems to be sufficient as a risk assessment.

Lack of Risk Assessment: Uncertainty Due to a Lack of Full Safety

In Alpharma no SCAN opinion was available on the specific substance and in *Solvay* a risk assessment was not performed by any EU expert body or Member State competent authority. In Alpharma, the Court held that the available data was too scarce to conduct a risk assessment of the possible risks to human health as a result of using bacitracin zinc.[47] As Laduer (2003, p. 1456) also observes:

> [d]espite the fact that, in Alpharma, SCAN had not even been consulted at all, the Court was of the opinion that the Council and Commission had been sufficiently informed by the opinion given in the other procedure concerning Pfizer.

Such an argumentation is all the more surprising given that, in this particular opinion, SCAN alleged there to have been too much uncertainty to justify a risk assessment (SCAN 1998; Van Asselt and Vos 2006). The Court nevertheless ruled that the Commission did not err when it failed to perform a risk assessment of bacitracin zinc and did not request an opinion of SCAN. The Court also found that the Commission did not have to review the other materials at hand to ascertain whether or not they met EU standards.

In *Solvay*, a similar pattern may be observed. Nifursol was not examined by the Committee for Veterinary Medicinal Products and SCAN held that there was not enough data available to perform a full risk assessment. As a result of this, SCAN stated that the safety of Nifursol could not be guaranteed (SCAN 2001), which is always the case in situations of uncertainty. In its opinion on nitrofurans in general – to which Nifursol belongs – SCAN concluded that it was never proven that certain similar characteristics of nitrofurans could be the cause of any health risks. This clearly indicates the problem arising in situations of uncertain risks: the risk cannot be proven nor refuted. Both the VMD and the Court however, still inferred from that same conclusion that there could still be a risk. It could be argued that they fall into the trap of equating uncertainty with risk.[48] According to the Court, the VMD statement of risk could 'not be undermined by the fact that […] *there were no studies available* of the various substances considered, apart from furazolidone and nitrofurazone'[49] (emphasis added). The Court held that the safety of Nifursol could not be proven to a sufficiently certain level.[50] Thus, the Court did not reprimand the Commission and the Council for the lack of a risk assessment. It may be argued that as a consequence the Court forced itself in the role of risk assessor, when it actually ought to check whether the risk manager performed the established risk assessment duties.

From the above it can also be concluded that the Court engages in another way to construct uncertainty next to the two patterns identified above: constructing uncertainty as contrasting opinions and constructing risk (and discarding uncertainty) by analogical handclapping. In the Solvay Case, the Court established uncertainty as the lack of full safety.

The procedural importance of scientific risk assessment as a prerequisite for precautionary action (Stokes 2008) combined with the trend of the European Courts to increase their involvement in the scientific debates (Alemanno 2008), has forced the European Courts to construct the means by which to legitimise the principle and to evaluate whether uncertainty is present. Given that uncertainty has been construed as scientific uncertainty, the Courts deal with scientific uncertainty when reviewing applications of the precautionary principle (Stokes 2008). According to Sandin (2006b), this has led to a situation where '[t]here is room for further speculation as to what level of evidence is required to trigger precaution' (p. 181).

Defining what constitutes 'enough scientific evidence' appears to be a very problematic issue for the Court (Vos 2004). Can the Court determine this? Perhaps equally as important is the question of whether the Court should even have to deal with such cases of scientific complexity. The Court frames its cases in such a way that it is compelled to enter into a scientific discussion. Although in is wording it prefers to see its role as one of limited review, in practice it is acting as a risk assessor. One could argue that the Court is asking the wrong questions. By doing so, it has forced itself to become a super-expert, a role it simply cannot live up to. Therefore, it is our view that the Court should focus more on reviewing the process. In all cases, critical questions ought to have been asked about how the Council and the Commission addressed the risk assessment obligation. This could in turn have stimulated more of a fundamental discussion on the assessment of uncertain risks. Accepting that in situations of uncertainty experts cannot provide certainty with regard to the seemingly simple question 'whether there is a risk': the question is more one of what could or should be expected from scientific evaluation in the context of precaution-based risk management and also, therefore, what role experts might play in that evaluation.

Old Data Combined With a Passive Review of Evidence

In 1999, Nifursol was the subject of new authorisation under Article 9h of Directive 70/524 for which Solvay applied. The Courts ruling against renewal of market authorisation was based on an assessment done by the veterinary medicinal products committee carried out between 1990 and 1995. The Court argued that a new scientific assessment of Nifursol was not necessary since the re-evaluation was administrative in nature.[51] The replacement of provisional authorisation for Nifursol was based on Article 9h of Directive 70/524. This Article provides that two conditions must be fulfilled. First, the identification notes and monographs concerning the additive must be submitted on time, and second, that they must be in accordance with the data upon which the original authorisation was based. Thus, the Court held both that Article 9h of Directive 70/524 concerned merely an administrative procedure and that the Commission had not erred in this regard. Therefore, Solvay could not rely on the ordinary rules of authorisation and re-examination[52] which would have required a new risk assessment. The Court then held that although the re-evaluation was administrative in nature, the information originally submitted in the authorisation procedure gave no reason to believe that Nifursol was safe.[53] The Commission did not examine whether new insights pertaining to risks were available and the Court subsequently disregarded this oversight. Furthermore, the Commission did not discuss the reason why the same information could lead to divergent outcomes: authorisation *versus* refusal of renewal of the authorisation. Due to the 'administrative nature' of the re-evaluation of Nifursol's authorisation, the data provided in the Solvay Case was old data. Apparently, the Court found this not to contradict the requirements for a risk assessment.

In the Pfizer Case, the Court ruled that precautionary measures must be based on a risk assessment and this risk assessment should be as thorough as possible based on the latest scientific evidence available.[54] In addition, a risk assessment should adhere to the principles of excellence, transparency and independence.[55] Vos (2004) observed in *Pfizer* and *Alpharma*, that '[t]he Court ... was not consistent in following its own criteria' (p. 192). The threshold of scientific evaluation necessary to trigger precautionary measures is very low in the Solvay case. The Court may have found that Article 9h of Directive 70/524 prescribes a purely administrative review, but is this desirable? When assessing the safety of a substance for present purposes, at least the data should be up to date. By allowing for a passive review, the Court ignores that the parties are obliged to attain and investigate new evidence and new insights, as is prescribed in the Commission's own guidelines for the precautionary principle (European Commission 2002a). Actions based on the precautionary principle should be subject to periodic review in order to determine whether the scientific understanding of the risk has improved (WRR 2009; Rogers 2011). Rogers also observed that the Court has not been very strong in upholding this precautionary criterion (Rogers 2011). In this manner, the Court fails to take the temporary character of the precautionary principle seriously.

The Commission also relied on old data in the Artegodan Case. On 9 March 2000 the Commission adopted Decision C(2000) 452 withdrawing the market authorisation for phentermine. It was held that 'the evaluation of the efficacy of phentermine had not changed since the CPMP opinion of 17 July 1996'.[56] Moreover, it was pointed out that 'the best available evidence for efficacy in long term use [...] came from the 1968 report by Dr. Munro and others'.[57] Nevertheless, in its final opinion of 31 August 1999 on phentermine, the CPMP recommended withdrawal of market authorisation. The scientific data considered by the Commission in 1999/2000 were completely identical to those of 1996,[58] but its risk management measures were diametrically opposed. For the substance amfepramone, the CPMP relied on a similar line of reasoning.[59]

The ruling of the Court in the Artegodan Case was very different compared to the preceding cases. The Court then only reviewed the consistency of the CPMP opinion and its legality. It did not – as it tried to do in the Pfizer, Alpharma and Solvay Cases – go into the merits of the arguments. The Court argued that:

> [t]he precautionary principle requires the suspension or withdrawal of a market authorisation where *new data* give rise to serious doubts as to either safety or the efficacy of the medicinal product in question[60] (emphasis added).

The new data needs to be 'solid and convincing evidence'.[61] The Court then continued to emphasise the importance of 'the progress of scientific knowledge'[62] and 'new discoveries'.[63] When addressing the CPMP's claim that the risks posed by the substances had not changed since 1996, the Court then continued its argumentation by stating that:

> [i]n mentioning in its scientific conclusions on amfepramone and on phentermine that the risk of cardiac valve disorder could not be ruled out, the CPMP was merely stating that *no evidence could be supplied to show that there was no such risk*. Moreover, its scientific conclusions explicitly stated that, for all the substances under consideration, there was *no solid evidence* to justify an assumption that their use increases the risk of cardiac valve disorders. In addition when carrying out its assessment of the benefit/risk balance of the substances in question, the CPMP weighed their purported lack of efficacy against *only those risks which had already been taken into consideration in 1996*[64] (emphasis added).

Additionally, the Court held that 'the CPMP opinions of 31 August 1999 and the contested decision […] are based on medical and scientific data […] which are *strictly identical* to those taken into consideration in 1996'[65] (emphasis added).

In the Artegodan ruling the Court did not accept a reinterpretation of the uncertainty observed before as the sole justification for invoking the precautionary principle. It argued that any judgment about uncertainty at a present time in order to make the case for withdrawal needs to be informed by new data. It explicitly argued that old data which has been used in previous assessments may not constitute a sufficient basis upon which to establish scientific uncertainty in the present. In adopting a more procedural approach, the Court was able to ask critical questions about the ways the risks and uncertainty were assessed.

The inconsistency in the Court's behaviour in *Artegodan* compared to the Solvay Case is striking. In *Artegodan*, the Court argued that the old data (on which an earlier decision on market approval was based) is neither enough to constitute solid evidence nor scientific uncertainty and that new evidence is required to legitimise withdrawal of market authorisation. In contrast, the Court held that the Commission did not err in just using old data to argue in favour of withdrawal. In that case, the Court accepted that the case was cast as purely administrative in nature.

Tensions and Patterns

The patterns that are visible when discussing the level of uncertainty required for precautionary action can be linked to the Court's dealing with the procedure and evaluation of the risk assessment. First, in several of the cases a risk assessment is lacking. There is either no SCAN opinion present or SCAN could not perform a full risk assessment. The Court did not reprimand the Commission for this in two of the three cases. It could be argued that the Court realised that there is a tension between uncertainty and the requirement of a full risk assessment, but if that was the case, in our view, the Court should have indicated a tension or friction in the regulatory regime. In that way, the Court ruling could have informed thinking about the regulatory regime surrounding the precautionary principle. Instead of taking this role however, the Court acted as a super risk assessor. Second, the Court found in *Solvay* that an administrative review is sufficient when deciding on precautionary measures. By not insisting on a new risk assessment of the substances, the Court also disregards the temporary character of the precautionary principle. Precautionary measures are supposed to be temporary in nature and should be subject to review when there is new scientific evidence. If such re-evaluation is not performed, precautionary measures run the risk of becoming permanent measures.

Overall, three tensions are apparent. There is a discrepancy between risk assessment as a procedure and the way in which the Court evaluates the lack of risk assessment. In addition, there is a discrepancy between the temporary nature of the precautionary principle and the lack of substantive review. Finally, there are inconsistencies in the way in which the Court reviews similar cases.

Conclusions and Discussion

The lack of a clear definition of the precautionary principle and the flexibility in its application challenges the Courts and the European Court of First Instance in particular. The Pfizer Case was the first case in which CFI explicitly reflected on the precautionary principle and reviewed precautionary measures in those terms. It has been argued that if the Court's ruled in that *Pfizer* – constructing uncertainty as the result of contrasting scientific opinions and as sufficient basis to legitimate a precautionary ban – were to serve as a precedent, this would render empty the precautionary principle and empty one. Therefore, we examined certain post-Pfizer case law in which the CFI ruled on the application of the precautionary principle: *Alpharma*, *Solvay* and *Artegodan*. Table 9.1 summarises these findings.

Table 9.1 Overview of Case Law

Case	Disputed Facts	Ruling	Patterns & Inconsistencies
Pfizer	The Commission imposed a ban on the use of virginiamycin as additive to animal feed. The possible risk to human health as a result of transferring antibiotic resistance from animals to humans through the use of virginiamycin.	The CFI ruled that there was a scientific basis for a possible link between the use of virginiamycin and developing resistance in humans. Thereby, it upheld the ban.	Uncertainty was constructed as contrasting scientific opinions. The Court assigned itself the role of risk assessor. Product framing: the Court attempts to examine the scientific evidence.
Alpharma	The Commission imposed a ban on the use of bacitracin zinc as additive to animal feed. The possible risk to human health as a result of transferring antibiotic resistance from animals to humans through the use of bacitracin zinc.	The Court ruled in favour of the Commission and upheld the ban of bacitracin zinc.	*Pfizer* precedent: Uncertainty constructed as contrasting scientific opinions. Lack of risk assessment: No risk assessment of bacitracin zinc. The Court as risk assessor. Analogy: SCAN opinion on virginiamycin used as scientific basis for uncertainty.
Solvay	The Commission withdrew market authorisation for Nifursol. The possible risk of genotoxicity (leads to mutations in cells) of Nifursol.	The CFI ruled that the safety of Nifursol could not be guaranteed and upheld the withdrawal of market authorisation.	Lack of risk assessment: Old data which did not include Nifursol. SCAN: full risk assessment not possible. Analogy: General claim on characteristic antibiotics. Uncertainty as lack of safety. Passive review: Temporary nature of precautionary principle discarded. Product framing.

Table 9.1 (continued)

Case	Disputed Facts	Ruling	Patterns & Inconsistencies
Artegodan	The Commission withdrew market authorisation for five substances used in the treatment of obesity. The possibility of increased risk for Primary Pulmonary Hypertension (PPH) (high blood pressure leading to heart failure).	The Court ruled in favour of the applicants, finding the Commission to have erred since it had not performed a proper risk assessment.	The Court adapted a more stringent procedural approach: Old data is not sufficient to establish uncertainty. Progress and new scientific data needed: uncertainty cannot always remain. Process framing: the Court did not engage in the scientific debate, but reviewed the process.

Our analysis reveals that the Pfizer Case indeed set a precedent. Also in Alpharma, the Court constructed uncertainy as the result of scientific dissensus. Since scientists disagree in virtually all uncertain risk cases, by implication the precautionary principle may be applied whenever any single qualified expert disagrees with his peers: an unfortunate precedent which undermines the meaningful use of the precuationary principle as a risk management too.

Our analysis further reveals that there are other problematic patterns in how the Court deals with uncertainty and how it reviews measures that are defended with reference to the precautionary principle:

- Constructing uncertainty as contrasting scientific opinions (*Pfizer* precedent);
- Constructing risk (and discarding uncertainty) by analogical handclapping;
- Constructing uncertainty as the absence of full safety;
- Engaging in risk assessment (instead of reviewing whether and how the Commission and the Council satisfied the risk assessment duties);
- Ignoring the temporary nature of precautionary measures and the associated duty to engage in re-examination (instead of accepting the concept of administrative review and old data);
- Inconsistencies in dealing with uncertainty and the relationship between uncertainty and precautionary measures between cases.

It would be wrong to read our analysis purely as a criticism of the Court's behaviour. We also level accusations at the Court of it having created deeper tensions and frictions in the regulatory regime: the uncertainty paradox; the debatable role of risk assessment in situations of uncertainty; the tension between a product-based regulatory regime; and a political culture which prefers to debate technologies and risk on different levels as well as the tendency to equate uncertainty with risk.

Nevertheless, our analysis indicates that the European Courts lack a clear vision of how to rule on uncertainty and precaution. The Court seems to find it difficult to shape the relationship between scientific evidence, uncertainty and the precautionary principle. Through case law, the Court, on the one hand, has established the boundaries and rules governing the application of the precautionary principle. At the same time, it ignores these self-established boundaries and rules, allowing the Commission and the Council to do the same, with the result that previously established rules and boundaries become empty promises. The Commission's *Communication on the Precautionary Principle* emphasises the temporary nature of precautionary measures: they should be re-evaluated in light of scientific development. Rogers (2011), evaluating a series of cases before the European courts, including recent ones, argues that the Courts fail to uphold the temporary character of the principle when they do not review precautionary measures. He concludes that '[i]f the [C]ourts were to take account of this criterion in their judgments it would […] require the regulator to follow the rules set out in the Communication' (p. 481). The problematic tensions, observed in this contribution, arising in the way the Courts have dealt with uncertainty, old data and precaution, add to Rogers 2011 analysis.

The Pfizer Case could have been the starting point for the Court to hold itself and everyone else to high standards for reviewing precautionary measures. That it failed to do so is a missed opportunity (Vos 2004). It also suggests that more case law would probably not be the way to develop a vision and to dissolve the inconsistencies in rulings on precautionary measures. As a way forward, we envisage two different approaches. In the first approach – to be undertaken urgently – the Court needs to develop or demand a vision in order to 'save' the precautionary principle as a tool for risk management. This vision could become concrete, for example in a new Commission *Communication on the Precautionary Principle*. The second approach follows Millstone's argument that it is inherently problematic to set a risk assessment as a pre-requisite to the application of the precautionary principle (Millstone 14 April 2010; WRR 2009). Risk assessment and the role of experts in regulatory procedures should be reviewed and rethought.

If the problematic patterns within the Pfizer, Alpharma, Solvay and Artegodan Cases and the inconsistencies within and between them set precedents, this will lead to serious differences and more questionable rulings. This will in turn damage the attempt to organise responsible decision-making in view of (uncertain) risks, (uncertain) benefits and (uncertain) trade-offs.

Notes

The authors would like to thank Ellen Vos, Ragnar Löfstedt and Alberto Alemanno for their helpful input.

1 CFI Case T-74/00, Artegodan GmbH *versus* Commission of the European Communities [2002], at para. 184.

2 'Uncertain risks' is used here as a shorthand for risks to which Renn *et al.* (2010) and Van Asselt and Renn (2011) refer as complex, uncertain and/or ambiguous risks. Key to risk typologies is simple, calculable risks on the one hand, and other risks referred to here as 'uncertain risks' on the other hand.

3 CFI, Case T-13/99, Pfizer Animal Health SA *v.* Council of the European Union [2002], Judgment of 11 September 2002, OJ 2002 C 289/21.

4 CFI, Case T-13/99, *supra* note 3, paras. 29–31.

5 CFI, Case T-70/99, Alpharma Inc. *v.* Council of the European Union [2002], Judgment of the CFI of September 2002, OJ 2002 C 289/21.

6 CFI, Case T-13/99, *supra* note 3, para. 41.

7 In this case, both the risks of transfer of virgianiamycin from animals to humans and the associated risks of development of antibiotic resistance in humans as well as the transfer of antibiotic resistance from animals to humans are discussed.

8 CFI, Case T-13/99, *supra* note 3, para. 393.

9 *Ibidem.*

10 CFI, Case T-13/99, *supra* note 3, para. 394.

11 CFI, Case T-13/99, *supra* note 3, para. 389.

12 CFI, Case T-13/99, *supra* note 3, para. 369.

13 CFI, Case T-70/99, *supra* note 5, paras. 43–44.

14 CFI, Case T-70/99, *supra* note 5, paras. 45–53.

15 CFI, Case T-392/02, Solvay Pharmaceuticals BV *v.* Council of the European Union [2003], Judgment of 21 October 2003, OJ 2003 C7/35, paras. 27–28.

16 CFI, Case T-392/02, *supra* note 14, paras. 41, 145.

17 CFI, Case T-392/02. *supra* note 14, paras. 26–57.

18 CFI, Case T-392/02, *supra* note 14, paras. 147, 164.

19 CFI, Case T-74/00, *supra* note 1, para. 14.

20 CFI, Case T-74/00, *supra* note 1, para. 24.

21 CFI, Case T-74/00, *supra* note 1, para. 38.

22 CFI, Case T-74/00, *supra* note 1, paras. 61–70.

23 CFI, Case T-74/00, *supra* note 1, paras. 220–21.

24 CFI, Case T-13/99, *supra* note 3, para. 393.

25 CFI, Case T-70/99, *supra* note 5, paras. 30–42.

26 CFI, Case T-70/99, *supra* note 5, para. 37.

27 CFI, Case T-70/99, *supra* note 5, para. 41.

28 CFI, Case T-70/99, *supra* note 5, para. 150.

29 CFI, Case T-70/99, *supra* note 5, para. 167.

30 CFI, Case T-70/99, *supra* note 5, para. 299.

31 CFI, Case T-70/99, *supra* note 5, para. 307.

32 CFI, Case T-70/99, *supra* note 5, para. 306.

33 It should be noted that in the SCAN opinion it was actually argued that there was too much uncertainty to do a risk assessment, so no transfer mechanisms were established, only hypothesised. Based on the available scientific knowledge, SCAN concluded that there was no immediate risk. It was already quite doubtful that in the Pfizer case the SCAN opinion is repeatedly referred to as risk assessment. But it is even more surprising that it is used in analogical reasoning as if the SCAN opinion on virginiamycin was a risk assessment in which transfer mechanisms were established.

34 CFI, Case T-392/02, *supra* note 14, para. 141.

35 CFI, Case T-392/02, *supra* note 14, paras. 41, 145.

36 *Ibidem.*

37 CFI, Case T-392/02, *supra* note 14, para. 139.

38 *Ibidem.*

39 CFI, Case T-392/02, *supra* note 14, para. 141.

40 CFI, Case T-392/02, *supra* note 14, para. 143.

41 CFI, Case T-392/02, *supra* note 14, para. 129.
42 Jasanoff (2005) distinguishes between three ways in which technology can be conceptualised, 'as a technoscientific *process*, as a stream of *products* or as a *program* of governance and control' (p. 45).
43 To which Jasanoff (*Designs on Nature*) refers as process-base.
44 CFI, Case T-13/99, *supra* note 3, para. 144.
45 Order of the Court of First Instance, 14 December, 2005. Arizona Chemical and Others v. Commission, CFI, Case T-369/03, para. 85.
46 Judgment of the Court, 9 September 2003, in ECJ, Case C-236/01. Monsanto Agriocultura Italia SpA v. Presidenza del Consiglio dei Ministri, para. 131.
47 CFI, Case T-70/99, *supra* note 5, para. 44.
48 For a more extensive discussion of this pitfall see: Van Asselt and Vos (2008).
49 CFI, Case T-392/02, *supra* note 14, para. 144.
50 CFI, Case T-392/02, *supra* note 14, paras. 147, 164.
51 CFI, Case T-392/02, *supra* note 14, paras. 116–33.
52 CFI, Case T-392/02, *supra* note 14, paras. 109–10.
53 CFI, Case T-392/02, *supra* note 14, para. 116.
54 CFI, Case T-13/99, *supra* note 3, para. 162.
55 CFI, Case T-13/99, *supra* note 3, para. 159.
56 CFI, Case T-74/00, *supra* note 1, para. 48.
57 CFI, Case T-74/00, *supra* note 1, para. 53.
58 CFI, Case T-74/00, *supra* note 1, para. 204.
59 CFI, Case T-74/00, *supra* note 1, para. 54.
60 CFI, Case T-74/00, *supra* note 1, para. 192.
61 *Ibidem.*
62 CFI, Case T-74/00, *supra* note 1, para. 193.
63 *Ibidem.*
64 CFI, Case T-74/00, *supra* note 1, para. 203.
65 CFI, Case T-74/00, *supra* note 1, para. 204.

Bibliography

Alemanno, A. (2007) 'The Shaping of the Precautionary Principle by European Courts. From Scientific Uncertainty to Legal Certainty', in L. Cuocolo and L. Luparia (eds) *Valori Costituzionale e Nuovo Politiche Del Diritto*, Cahiers Européens, Halley, Bocconi Legal Studies Research Paper 1007404.

——(2008) 'EU Risk Regulation and Science: The Role of Experts in Decision-Making and Judicial Review', in E. Vos (ed.) *European Risk Governance. Its Science, its Inclusiveness and its Effectiveness*, Mannheim: Connex.

Case Law Analysis (2003) 'Fleshing Out the Precautionary Principle by the Court of First Instance', *Journal of Environmental Law*, 15: 372–405.

Da Cruz Vilaça, J.L. (2004) 'The Precautionary Principle in EC Law', *European Public Law*, 10: 369–406.

De Sadeleer, N. (2007) 'The Precautionary Principle in European Community Health and Environment Law: Sword or Shield for the Nordic Countries?', in N. De Sadeleer (ed.) *Implementing the Precautionary Principle. Approaches from the Nordic Countries, EU and USA*, London: Earthscan.

European Commission (2002a) *Communication from the Commission on the Precautionary Principle*, 2 February 2002, 1–29.

——(2002b). 'Commission Regulation 178/2002 of the European Parliament and the Council of 28 January 2002 laying down the general principles and requirements of food law, establishing the European Food Safety Authority and laying down procedures in matters of food safety', *OJ* 2002 L 31/1.

Forrester, I. and Hanekamp, J.C. (2006) 'Precaution, Science and Jurisprudence: A Test Case', *Journal of Risk Research*, 9: 297–311.

Hahn, R.W. and Sunstein, C.R. (2005) 'The Precautionary Principle as a Basis for Decision Making', *The Economists' Voice*, 2: 1–9.

Jasanoff, S. (2005) *Designs on Nature. Science and democracy in Europe and the United States*, Princeton: Princeton University Press.

Kaiser, M. (1997) 'The Precautionary Principle and its Implications for Science – Introduction', *Foundations of Science*, 2: 201–5.

Ladeur, K.H. (2003) 'The Introduction of the Precautionary Principle into EU Law: A Pyrrhic Victory for Environmental and Public Health Law? Decision-Making Under Conditions of Complexity of Multi-Level Political Systems', *Common Market Law Review*, 40: 1455–79.

MedlinePlus (2011) 'Pulmonary Hypertension', 28 March 2011, available online: <http://www.nlm.nih.gov/medlineplus/ency/article/000112.htm> (accessed 27 April 2011).

Millstone, E. (14 April 2010) 'European Integration Between Trade and Non-Trade Conference', Maastricht University.

Percival, R.V. (2005) 'Who's Afraid of the Precautionary Principle?', *Pace Environmental Law Review*, 23: 1–76.

Rogers, M.D. (2011) 'Risk Management and the Record of the Precautionary Principle in EU Case Law', *Journal of Risk Research*, 14: 467–84.

Sandin, P. (2006a) 'The Precautionary Principle and Food Safety', *Journal of Consumer Protection and Food Safety*, 1: 2–4.

——(2006b) 'A Paradox out of Context: Harris and Hold on the Precautionary Principle', *Cambridge Quarterly of Healthcare Ethics*, 1: 175–83.

SCAN (1998) 'Opinion of the Scientific Committee for animal Nutrition on the immediate and longer-term risk to the value of Streptogramins in Human Medicine posed by the use of Virginiamycin as an animal growth promoter' (produced at the request of the Commission in response to the action taken by Denmark under a safeguard clause to ban Virginiamycin as feed additive), 10 July 1998. available online: <http://ec.europa.eu/food/fs/sc/scan/out14_en.html> (accessed 6 April 2011).

——(2001) 'Opinion on the Safety of Use of Nifursol in Feedstuffs for Turkeys', 11 October 2001, 1–15.

Stokes, E. (2008) 'The EC Courts' Contribution to Refining the Parameters of Precaution', *Journal of Risk Research*, 11: 491–507.

Tetlock, P.E. (2005) *Expert political judgement: How good is it? How can we know?* Princeton: Princeton University Press.

UN Economic and Social Council, *Rio Declaration on Environment and Development: application and implementation*, 10 February 1997, available online: <http://www.un.org/esa/documents/ecosoc/cn17/1997/ecn171997–98.htm> (accessed 29 March 2011).

Van Asselt, M.B.A., Faas, A., van der Molen, F. and Veenman, S.A. (eds) (2010) *Out of Sight: Exploring Futures for Policymaking* (in Dutch), Amsterdam University Press, WRR Verkenning 24.

Van Asselt, M.B.A. and Vos, E. (2006) 'The Precautionary Principle and the Uncertainty Paradox', *Journal of Risk Research*, 9: 313–36.

——(2008) 'Wrestling with Uncertain Risks: EU Regulation of GMOs and the Uncertainty Paradox', *Journal of Risk Research*, 11: 281–300.

Van Calster, G. and Lee, M. (2003) 'European Case Law Report July – November 2002', *RECIEL*, 12: 109–14.

Vos, E. (2004) 'Antibiotics, the Precautionary Principle and the Court of First Instance', *Maastricht Journal of European and Comparative Law*, 11: 187–200.

WRR (2009) *Uncertain Safety. Allocating Responsibility for Safety*, Amsterdam: Amsterdam University Press.

10 The Scientification of Participation

Frank Rodrigues

Introduction: Risk Analysis Versus Risk Governance

To suggest that a sound policy solution to potential risks requires understanding and responsiveness is unlikely to surprise or illuminate the reader. The discussion becomes more intriguing though if we ask what such understanding and responsiveness should look like. Broadly speaking, the question has had two answers built around two distinguishable, theoretical approaches. For convenience, they may be labelled *risk analysis* and *risk governance*. Risk analysis – at times described as traditional risk governance – involves risk assessment, risk management and risk communication. Essentially, the focus has been on an analytical and technical understanding based on scientific method: quantitative risk assessments are therefore preferred; risk management is reduced to interpreting quantitative results and controlling their impact, and risk communication is an exercise aimed at bridging the tensions that may arise between risk assessors and the public (Aven and Renn 2010: 50–52). The public is the passive recipient of the risk assessor's solutions and is, by implication, part of the matrix of circumstances that needs to be risk-managed. Once scientific understanding through risk assessment is achieved, what counts as proper responsiveness – broadly framed as a blend of risk management and risk communication – follows. One might say that, in the case of risk analysis, understanding acts as a necessary and sufficient condition for responsiveness. Once determined, scientific truth – mostly under the control of 'experts' – will deliver *ipso facto* an acceptable response to potentially risky events.[1]

A more recent answer to the question of how we should conceive of the understanding of and our responsiveness to risk, has morphed risk analysis into the notion of risk governance. In the latter, particularly in relation to risks believed to be potentially systemic, i.e. affecting the whole system that humans live in, the socio-cultural context has to be incorporated into the 'factual' dimensions of risk established by risk assessors. For risk governance, understanding and responsiveness additionally involve the varying values and perceptions of pluralist societies (Aven and Renn 2010: 53) and the co-ordination mechanisms such as markets, incentives or self-imposed norms [...] including concern assessment and explicit discussion of stakeholder participation (Renn 2008: 9). Another way of making the point is to say that this approach tends to make understanding and responsiveness interactive: in fact, while understanding may be a necessary condition for responsiveness, it can never be sufficient as the latter requires a connection to lived human values and concerns. The risk governance approach focuses on the prospect that while understanding may involve the use of an analytical reason, responsiveness demands consideration of social, and possibly ethical, values. Indeed, one may go a step further and say that a technical risk assessment if not wrong, will at best be incomplete, if it is not informed by potential risk management concerns and vice versa.[2] Arguably, such an approach is beneficial where a given risk problem is complex, uncertain or ambiguous, irrespective of whether such a risk is considered to be systemic or not.

Traditional risk analysis may be said to be rooted in a calculative reason, which treats risks as essentially identifiable and measurable, a technique initially prevalent in the assessment of commercial and financial risks and then gradually adapted in the nineteenth century for the purposes of controlling the spread of illness, levels of poverty and rates of crime (Wilkinson 2010: 18). In its purest form it yields a technocratic model of policy making, addressing a wide range of human concerns, where scientists are seen to be the best judges of risk assessment and management. Milder incarnations of traditional risk analysis deliver a decisionistic model that divides policy making methodologically and substantively into the scientific assessment of risk and a risk management which also takes ethical and political concerns into account (Renn 2008: 11). The emphasis on a scientific, technical analysis of risk 'provides society with a narrow definition of undesirable effects and confines possibilities to numerical probabilities [...] derived from experiments, models, expert judgements, and so on' (Renn 2008: 42).

In contrast, risk governance does not, or at least is not supposed to, privilege science nor does it methodologically or substantively separate out risk assessment from risk management. Instead, science, politics, economic actors and representatives of civil society are invited to play a role in both assessment and management (Renn 2008: 11). It replaces calculative reason with a social sciences perspective and expands the horizon of risk outcomes by referring to socially constructed or socially mediated realities (Renn 2008: 42). Access to such realities is provided through the public participation of all relevant stakeholders and groups in 'the process of framing the [risk] problem, generating options, evaluating options and coming to a joint conclusion' (Renn 2008: 274). This process is a mutual, on-going dialogue between risk assessment and risk management, risk generators, regulators and those affected by the risk, and is not a one-off event (IRGC 2009: 62).

Participation is Central to Risk Governance

There is more than an echo of the words and ideals of participatory democracy (self-determination and self-rule) in the way in which risk governance through public participation is described. Indeed, it is very much a case of the word being made flesh. Public involvement is seen to be morally and functionally integral to the two key ideas of democracy: popular sovereignty and political equality (Webler and Renn 1995: 21). Similarly, in risk governance public participation is required for ethical, political and epistemological reasons (Andersson 2008: 141–42). It allows issues of fairness to be addressed, (re)builds public trust and creates a context in which the public may be heard. 'It leads to a sense of ownership of the outcome and helps reassure the public ... that regulators and policy makers have their best interests at heart' (Löfstedt and Van Asselt 2008: 82). Writers have been keen to stress that public participation should find a place in both risk assessment and risk management, in our understanding of risk and our responsiveness to it, in how we determine what it is, and in how we allow ourselves to be ruled in order to deal with it (Löfstedt and Van Asselt 2008).[3]

So far, so good and much in tune with the current *Weltanschauung* where the word *democracy* is almost always accorded a positive valence (Agamben *et al.* 2011: vii). However, if risk governance is the place where state and non-state actors come together to define and apply risk regulations or norms through a process of participation, and if participation is the locus where understanding and responsiveness are blended and where risk assessment and risk management are integrated, then participation is also potentially the medium through which laws are both created and enforced or at the very least, become enforceable. The process of participation in risk governance in effect provides the content of the law alongside the means of its application. When most successful, it identifies norms and disarms any potential for disobedience: it is difficult to advise non-compliance with norms in whose making you have participated. Once technical understanding is in place, participation becomes the necessary and sufficient condition for understanding and responding to risks. The public becomes legislator and law-enforcer. Seen as such, public participation carries an implicit, and perhaps contradictory, subtext suggesting that the doctrine of the separation of legislative and executive functions of government, so dear to so many forms of democracy to prevent abuses of power, is *passé*. Self-determination and self-rule parade a logic that excludes self-abuse. The exercise of participatory choices guarantees the lawful exercise of power.[4]

How and in what sense could public participation, as an essential and seemingly innocent component of risk governance, imply such a dramatic message? There are several points that could be lined up to provide an answer. First, public participation as an ideal form of citizenship in Western democracies is broadly accepted. Its credentials appear to be impeccable. While participation can get messy in reality, it feels like a jolly good aspiration and to quibble with the fact only enhances the value. Embodying, possibly, the essence of democracy, the affirmative, normative force of participation is rarely questioned. Should we question it though? To what extent is public participation more of an ideology than an ideal? Is it a warm and cosy way of securing our political consent? If ideology is identified as the place where particular choices seem to be obviously correct and the only ones to choose, public participation is a good candidate for suspicion.[5]

Secondly, participation is inserted into risk governance discourse as a means of introducing a social perspective into the fact–value mix required for our understanding of and responsiveness to risks, particularly those that are uncertain or ambiguous. While scientific reasoning is meant to provide the facts, the societal context is meant to provide the values. However – and this is the main focus of this paper – we shall see that at a conceptual level, methods of participation are typically modelled on scientific method. Norms and values are slippery things: they feel safer when we can deal with them as fact. This conceptual glide of values to facts – referred to in this paper as the scientification of participation – deftly turns moral imperatives into objective social realities and marks a useful strategic move in the production of ideology.

Thirdly, participation has clearly had successes in dealing with localised issues even where the attendant risks have been complex, uncertain or ambiguous.[6] There is a temptation to extrapolate from solving for *these* risks to solving for *all* risks, even where a risk is potentially systemic. Systemic risks affect the systems on which society depends and connect natural events to economic, social and technological developments and policy driven actions, both at the domestic and the international level (Aven and Renn 2010: 18). While it is acknowledged that the investigation of systemic risks requires a move beyond the usual agent–consequence analysis to focus on interdependencies and spill-overs between risk clusters (Aven and Renn 2010: 19), the underlying belief is that the move from localised risks to systemic ones simply requires that a similar participation-centric risk governance approach be adopted, only this time bigger and broader (Aven and Renn 2010: 20, 49–66). We should cast our net more widely and more inclusively to ensure that all data and all stakeholders are captured. The problem with this framework is that it expects that participants in risk deliberations will approach systemic risks with shared assumptions and shared values, i.e. that the deliberations will be contained within a system. This can be seen as a further step in the value-to-fact glide noted above: not only are values putative facts but we can all find an objective haven where we will find we share them.[7] Such a stance becomes harder to maintain however if we move through the spectrum of systemic risks from the biological to the manmade: it is easier to find shared assumptions in relation to tackling systemic issues connected to tsunamis *versus* food safety. Food safety issues in turn have had to jostle less to find room on the risk agenda when compared to climate change. And a common response to climate change has been more tractable than the disparate voices heard in relation to the 2008 global financial crisis.[8]

Why do we have this unquestioning consensus around the value of participation as a regulatory ideal in risk governance? Why the move to dress participatory outcomes in scientific clothing and cover them, even in the context of systemic risks, with a cloak of objectivity? These questions lead us to the fourth point: participation with its twin legislative and executive functions, its role as law-maker and enforcer can act as a system of governmental control where ideological outcomes may be masked as social realities or socially inevitable. Participation has become widespread because many see it 'as necessary for system stability [...] reflected in the common claim that participation can enhance the responsiveness and legitimacy of public institutions' (Webler and Renn 1995: 23). Theoretical discussions of participation focus less on its putative executive function and more on its role in legislative decision making.[9] Less attention is therefore paid to how participation in risk governance performs its executive functions, to how a process that is meant to empower can actually disempower and also to how participation may be a risky business in and of itself.[10] This paper concludes with a brief examination of how the concept of 'governmentality' can shed light on some of these issues and the manner in which the law-enforcement aspect of participation can turn responsiveness into control.

The Scientification of Values

Set out below is a description of how participation becomes scientified. The argument begins with a discussion of how participation is prevalent in prescribed responses to risk, particularly uncertain risks. The prevalence occurs because in cases of uncertainty, scientific method may find it hard to separate fact from values or real events from hypotheses. Participation appears to acknowledge this uncertainty and positions it within a socio-ethical framework. To perform this task however, participatory methods have used an appeal to a super-objective rationality in order to harmonise scientific and social concerns. However, such harmony can be hard to find in the case of systemic risks if shared assumptions are unavailable. Even in the case of non-systemic risks, it may be hard to integrate the concerns of the whole gamut of relevant stakeholders.

Uncertain risks nevertheless help show up the limitations of risk analysis or traditional risk governance. The traditional model, based on a scientific paradigm, embraces a unitary, top-down approach to regulation which favours the production of certainty by using defined procedures to reduce complex phenomena into simpler constituents. Uncertain risks disrupt this unitary model, and participatory methods – central to risk governance approaches – seem to provide pluralistic regulatory norms which might better suit complex events. Arguably, there is a potential circularity of legitimacy in participation itself as it needs a legitimising norm to designate it as a legitimate norm-generating mechanism. The sense of circularity can be linked to an impetus to depoliticise the contest of values which participation is supposed to address by scientifying participation, i.e. by securing its legitimacy by an appeal to a meta-criterion, a super-objective rationality beyond the rationality of science. Writers who adopt this approach can be seen to advocate a form of scientific participation that is in thrall to the same 'philosophical imaginary' as the scientific paradigm itself. Changing the model of science from, say, the classical Newtonian model – which looks at universalising laws – to a more expanded, pluralistic one will not be sufficient to knit together the disjunction between scientific and social views.

Invoking the meta-criterion of super-objective rationality, can make participatory outcomes appear to be incontestable, wholly legitimised and outside political territory. They are then positioned to act as systems of control, a form of governmentality where the enforcement of the norm is implicit in its creation. Avoiding such potentially unconscious acquiescence requires participation to retain its social links, ground itself in the spheres of social will and social values and reject the hope of being sanctified by science.

Participation as a Response to Uncertain Risks

Public participation presents a promise of individual engagement accompanied by the warmth of collective solidarity. It seems to create a space for conflicting values to be synthesised and possibly provides – where a stable synthesis of conflict is impossible – at least partial foundations on which more controversial decisions may be socially embedded. Participation offers itself up as a political right which, when properly exercised, delivers transparent and legitimate decisions, regulations and law. What constitutes its proper exercise may be up for debate but its place on the democratic agenda is increasingly indisputable. As such, it may be assimilated by both the mainstream right and the mainstream left. In the UK, surely, the call of the current right-of-centre Government for the Big Society to replace the Big and Bad State is little more than a call to all citizens to participate with a vengeance. A call echoed by what was once Britain's New Left, where individual citizens are enjoined to inculcate active citizenship values in every community, from schooling upwards (Blunkett 2011: 4). This happy unity should come as no surprise: all good governments aim to give us what we really, really want. And participation in politics is the governmental hotline that transmits our wish-lists.

The tags of social embeddedness, transparency, legitimacy and democracy have also helped to assure participation a place in theoretical discussions on the regulation of uncertain risks.[11] Unlike uncertain risks, risks that are certain are considered to be those capable of being objectively and scientifically measured, modelled and assessed. Statistics show that drinking above a certain amount affects our ability to drive safely. This risk is therefore a scientifically determined 'known-known'. In consequence, we have in place mostly uncontroversial regulations specifying the maximum permissible drink-limit for drivers. Uncertain risks, in contrast, are not as neatly resolved. They arise because science has less confidence that it can objectively describe the outcomes of, or appropriate responses to, particular risky events. At best science may offer a choice of theories relating to its risk assessment or risk management, i.e. to its understanding of the event creating the risk or its regulating remedy. It is, however, unable to dish up its analysis of the risk-event as fact. One such risk-event is bovine tuberculosis in the UK where a controversial cull of badgers is being contemplated as a potential remedy (Defra 2010). Critics of the remedial measure warn, however, that it risks increasing the spread of the disease as infected badgers disperse over a wider area to escape the cull (Pickard 2011). Neither of the two contradictory theories of the impact of the cull can lay claim to the facts. Each, instead, could be said to have specifiable origins in a particular set of values.

Where scientific fact is hard to come by, regulation guided by scientific experts carries little weight. Regulatory theory has hence cast its net more widely in an attempt to find alternative benchmarks. A current response is to use a more pluralistic model involving participation as a means to achieve both substantive regulatory decisions and a guarantee of their legitimacy. Participation in this form embeds science within an ethical and social framework. While it might not eliminate uncertainty, it may at least legitimise or humanise it. Or so it would seem.

How Different is Participation as a Methodology?

Does participation truly provide a different methodology to aid our understanding of and responsiveness to risk? Not always, as the aim of using participation to provide legitimacy for regulation typically results in an appeal to a super-objective rationality capable of finding its way through the competing rationalities of an objective science and more subjective social discourses. This super-objective rationality is cast as a meta-criterion lying beyond, and acting as a bridge between, the scientific criteria that allow us to assess the claims of science and the social criteria which perform a similar task for social discourse. Hence, while participation is meant to provide the entire gamut of perspectives on the risky event, the meta-criterion of super-objective rationality allows the perspectives to be knitted together as a synoptic whole. We give up our partial points of view and see the world *sub specie aerterni-tatis*. Sometimes the meta-criterion is described substantively in terms, for example, of the common good; sometimes, more formally, as objective, deliberative procedures able to deliver coherent regulatory responses by the satisfactory and on-going resolution of conflicts.

Despite its attempt, then, to bridge the two spheres of the scientific and the social, the meta-criterion is ultimately modelled along the lines of a sci-entified system whose universal laws, connecting science to the social, can be unpicked with the application of sufficient diligence. This is unsurprising: the regulatory legitimacy that participatory frameworks seek to underpin is itself modelled on scientific legitimacy, i.e. on truths generated by the types of deductive or inductive reasoning used in science. Less attention is paid to a regulatory legitimacy based on social bonds (or the lack of them). This too is not wholly unexpected: a regulatory legitimacy that is socially derived, i.e. by a jostling of various social and subjective interests against each other and the resultant distribution of power, is modelled on political legitimacy and essentially re-introduces politics into regulation. This is something that regulatory theorists are minded to avoid. After all, regulation is meant to be fair and objective. The lobbyists should not win and the scientification of participation may help us believe that they have not.

Participation and Systemic Risks

Another reason for scepticism of the ability of participation to resolve the regulatory dilemma posed by uncertain risks relates to the extent to which it humanises the regulatory response. Undoubtedly, participation has had its successes. It has taught us to understand that one size does not fit all and that often the correct rule will not be universal but, instead, context-dependent. Such successes though, shine through best when they relate to risks within a system. An example of such a risk would be consultation by a local authority as to the best traffic management system that should be used at a busy local junction. In contrast, where risks are uncertain *and* systemic, finding context-dependent solutions through participation is almost a contradiction in terms. The system is the context and no amount of participating will change that.[12] A good example of the type of systemic risk in question is the current financial crisis. Responses from within the system have demanded better financial regulation over a wide array of issues ranging from living wills for banks to regulating bonuses for individual bankers. The concern though of the recent series of protests and city occupations across the globe is that the proposed regulatory changes are window-dressing and just an attempt by a vested interest to carry on doing business as usual. There is little belief that participating in regulatory reform, without addressing the need for political change, will provide solutions to the injustices perceived to be the issue. The assessment of those looking for solutions outside the system have suggested that the financial crisis is in fact a crisis of capitalism demanding a response that changes the socio-political and economic system itself.[13]

Where systemic risks are concerned, pursuing regulatory legitimacy through participation is to chase rainbows. Worse, there is a risk that participation in these circumstances, rather than humanising the regulatory response, serves to provide a super-objective cloak for sectarian interests. The scientification of participation can make us lose sight of the prospect that systemic risks might require a political response and not a regulatory one. One could therefore argue that scientification can in effect function as a means of displacing politics by institutionalising participation for a regulatory end. The paradox, and perhaps more cynically the hope, of institutionalisation is that the more developed an institution becomes, the more it becomes the province of experts and lobbyists and the less it becomes available for genuine political participation.

Even in the case of non-systemic risks, the scientification of participation works from the belief that it will be possible to cajole complex issues into a coherent regulatory response through the participation of interest groups, and that this will somehow provide a guarantee of the acceptability of such a response to the average citizen. That coherence will always be possible must be questioned. There will be a residual risk that, instead of a coherent regulatory policy, context-dependency produces a series of separate rules which are each applicable to separate and independent events. 'The problem is that the more actors, viewpoints, interests and values are included and thus represented in an arena, the more difficult it is to reach either a consensus or some other kind of joint agreement' (Renn 2005: 50). The need to secure regulatory legitimacy by reaching a synoptic view through participation can severely impair the ability of a regulator to produce substantive regulation.

The Components of the Risk Analysis Model

Uncertain risks show up the limits of traditional, unitary risk regulation models whether at the national, EU or global level. The calculative reason of traditional risk analysis is modelled on the scientific paradigm. The explanatory force of this paradigm is based on its ability to predict future outcomes with certainty. By and large, coherent explanations are achieved by applying methods of logical deduction or inductive experiment to a broad range of phenomena. Underlying these methods is the belief that phenomena are connected within a unified system. These connections may be hidden from us because of the complexity surrounding phenomena. The aim of scientific method is to reduce down the complexity to simpler constituents. At this simpler level, the connections between all phenomena in a unified system can be revealed. The scientific analysis of uncertain risks however shows that this may be easier said than done. There are choices that may be made in relation to how we understand the constituents of phenomena and their interconnections. One person's fact is another's theoretical reason.

The scientific paradigm may be seen to exert its influence on the main components of the traditional model of risk analysis. Hence the risk analysis model displays the following scientific or rationalising features.

A Top-down Approach to Regulation

At a macro-level, risk is identified, assessed and managed within a unitary framework of power. The regulator traces its legitimacy through a linear delegation of power from a sovereign to which it remains ultimately accountable. While power is delegated downwards, accountability directs itself upwards. Transparency is likely to follow the same trajectory as accountability: there is more visibility from on high. The sovereign has a bird's eye view of the system which enables her to control and hence unify it. One may also say that the greater the force of the upward trajectory of accountability, the greater the sovereign's control and the weaker the regulator's independence. The more the sovereign has control the more unified the system and the stronger the definitiveness of the regulatory response. Linear progressions of power are good at producing certain regulatory outcomes.

Certainty

Certain regulatory outcomes enhance the value of the rationalising impetus. Rationalising, in this context, and scientific reasoning go hand-in-hand. They typically work off deductive or inductive generalisations. Their success depends on a reductionism, the breaking down of complex wholes into unitary parts. Discrete constituents and defined procedures are the means to the end of regulatory certainty. Typically, this approach works best for regulation that is normative or prescriptive. Planning rules and procedures, for instance, are more likely to be capable of being tightly defined. There may be disputes about whether a heap constitutes a 'building' under planning rules but there will be little doubt surrounding the fact of the existence of that heap and that something is being done, or should be done, to it. Precautionary or remedial regulatory measures, in contrast, have fewer hard facts that can be reduced to their constituents. Witness the constantly revisited debate as to whether global warming and climate change are due to manmade or natural factors.

Discrete Constituents and Defined Procedures

Each constituent from which the regulator derives its power, or each constituent over which it holds power, is conceived of in a unitary fashion. The sovereign who provides the regulator with its clearly delineated mandate has, itself, a defined and definitive constitution. The identities of the citizens or civil bodies under the regulator's supervision can in turn be specified in a straightforward way. The State is able to unambiguously identify and facilitate regulatory institutions and functions. Potential hazards and the relevant regulatory response are capable of being modelled, generalised and simplified through defined procedures for the purposes of risk management. Responsiveness to a hazard is a matter of joining the dots between the rationally constructed units – sovereign, State, regulator, regulatory response and civil society. Joining the dots in this fashion preserves the linear delegation of power and ensures legitimacy. Another way to make the point is to say that the traditional model of regulation is based on a unified system of control which needs to exclude subjectivity or, at least, transform it into a set of objectively identifiable constituents in order to function. The philosophical imaginary at stake here is one where the regulator, the people regulated, the event inviting regulation and the interconnections between them can be reduced to same objective, constituent matter. The aspiration to this scientific objectivity provides transparency in regulation and secures a legitimate response.

Traditional risk analysis could therefore be said to employ a rationalising responsiveness based on scientific reasoning: it postulates a unified approach through which risks can be broken down into constituents for risk assessment and logically pieced together again for risk management. On this view, regulations ideally mirror or capture reality and this mirroring ensures that they are responsive and applicable to the real situations for which they are made. Arguably, this is only possible by making decisions as to what should or should not count as reality, as to what is objective fact and what is subjective fantasy. Scientific reductionism therefore, can produce certain regulatory outcomes by grounding substantive regulatory norms or formal regulatory principles within strictly circumscribed facts. These norms and principles inscribe the lines of power that thread together sovereign, State, regulator *et al.*, as beads on a string. Having and maintaining your precise position on the string preserves legitimacy. Interaction between each constituent part follows fixed prescriptions and participatory dialogue, to mediate between regulation and the regulated, occurs not as of right but by invitation. We will show that the influence of the scientific paradigm is replicated in the more recent risk governance model through the scientification of participation.

Disrupting the Risk Analysis Model

Uncertainty disrupts some, although not all, of the clear lines of the traditional regulatory paradigm. Uncertainties come in many shapes and sizes. Some may even be shapeless and impossible to measure. Some may have a social, political or economic impact and some may not. Some may be just about matters of fact; others may question the very basis on which we separate fact from non-fact. Some may have to do with the nature of the hazard itself, some with the adequacy of our regulatory response. The more an uncertainty moves from shaped to shapeless, from the physical world to the human, from facts to interpretations, from the objective to the subjective, the deeper its disruptive power. Moving along the uncertainty spectrum will move us though positions where science may be of help as the uncertainty arises only because we temporarily do not know – but will one day when we have a more comprehensive science – through to positions where science will be a hindrance because we permanently cannot know. Deploying scientific procedures in the latter case is inappropriate and will only serve to mask our epistemic inability and diminish our effectiveness. Distinguishing a temporary lack of knowledge from a permanent one will not always be a simple task: a temporary uncertainty as to whether an experiment is available to test the health risks of a particular GMO can be turned into a permanent dispute on the uncertainties and ambiguities of the experimental method utilised. Such disputes are even harder to resolve where uncertain, systemic risks are at issue. In such cases it will be impossible to specify a standpoint outside the system, which will provide the critical distance from which the facts of the system can be unearthed objectively. Systemic risk by definition requires us to evaluate it from within the system and ties us, whether we like it or not, to our subjective contexts.

Uncertainties then are potentially virulent and severe once contracted. They are disruptive to a point where we might find it impossible to say whether the best way of dealing with them is going to fit some mode of rationality or not. Even more fundamentally, it might put into question our powers to describe unambiguously the risk in question, and/or determine without controversy our regulatory response. Scientific validation will be in short supply and expert verification will not count for much. It is in this space where participation – the involvement of the public in a consultation to embed the regulatory response within a social and moral framework – is brought into play by theoreticians focusing on the regulation of uncertain risks.[14] Uncertainty in its various guises forces a transformation of the traditional model of regulation.

Participation and Pluralism

Compared to the high-definition traditional model of regulation, participatory models seem to have the blur of pluralism about them. At the very least, regulation is created by at least two actors: state and non-state. Rather than a unitary flow of regulation from top-down, the sources of regulatory input are dispersed and originate from a wide variety of constituents. Governmental regulation is transformed into regulatory governance, i.e. a combination of fixed law and informal procedures that provides responsiveness to risk, more particularly uncertain risk, through an on-going dynamic of influence and contestation between State institutions and civil society. Usually this on-going dynamic is described in deliberative terms and the participatory model is explicitly or implicitly[15] linked to deliberative democracy. With democracy comes the attendant legitimisation by the *demos* and, perhaps, confidence that regulation may be humanised and its responsiveness optimised.

When dealing with uncertain risks, participation begins where science ends. In the traditional model of regulation, science may be used to analyse the risk and prescribe the remedy. However, in cases of uncertainty science is unable to perform this function. At best it can offer various alternatives for us to choose from and participation is the method allowing us to make a choice. Participation may be seen as the process of creating procedural, formal, substantive and objective norms at the point at which objective, science-based rules are no longer available. It has to give voice to a plurality of interests and it is this giving voice which, in somewhat chicken-or-egg fashion, produces normative legitimacy. It will remain perpetually unresolvable as to which comes first: giving voice to produce legitimacy or whether the act of giving voice itself needs to be legitimised. Along with this circularity, and apart from the issue of legitimacy, additional concerns arise at the formal and substantive level.

Formally speaking, the concerns are two-fold. First, participation may be said to reproduce the empty space identified by various theorists as the metaphorical locus of democracy. Unlike in a monarchy, the people themselves do not govern, i.e. the seat of power is empty and occupied by appointees with limited accountability who can always choose to manipulate the democratic process for authoritarian ends. Secondly, giving voice to a plurality of interests may result in consensus, compromise or on-going conflict. To decide which of these norm-producing procedures should be preferred will itself be a normative act, thereby reproducing the circularity of legitimation at a formal level.

At a substantive level, the norms produced by participatory procedures may be seen to aggregate the interests of all participants – the 'will of all', to use Rousseau's terminology – or a negotiated derivation – the 'general will' – of all participants. The first will arguably be incapable of producing any norm at all while the latter will only be available at the cost of establishing a circular, formal procedure to generate it. Further, the process of negotiation even where deliberative will involve an element of horse-trading with an accompanying lack of transparency.

These concerns of circularity, legitimacy, accountability and transparency are arguably a result of understanding participation in terms of the same scientific paradigm which lies at the heart of the traditional, unitary models of risk analysis. The temptation is to channel participation into the production of certainties, to poke and prod a pluralistic framework until it will spit out a unity; to shift in effect the emphasis from responsiveness to the norm. The ability of participation to produce substantive rules for a variety of contexts is replaced with a prescriptive demand that it produces such rules for all contexts. Creating regulatory norms through a participatory process does sometimes work, broadly in cases where the risks being regulated are containable within a system and where uncertainty is a product of a temporary absence of scientific knowledge. Success in such cases can lead to participation being touted as a regulatory panacea for systemic risks where conclusive scientific knowledge will never be possible. In such cases, participation should be allowed to display its fundamentally political character. The mistake is to believe that participation through objective deliberative procedures will lose its clashing political colours and find, instead, a unifying regulatory remedy for systemic risk.

Participation and the Scientific Paradigm

Participation within the scientific paradigm – scientific participation – effectively appeals to a meta-criterion, i.e. a super-objective rationality beyond the objective rationality of science itself, in order to secure its legitimisation. This meta-criterion is needed because, where participation is invoked to deal with uncertain risks, something is required to bridge the gap between objective scientific criteria used in risk assessment and management and the subjective social criteria that will determine whether any such account of risk is acceptable and relevant. Whereas science produces its truth claims within a given methodological framework of deduction and induction, scientific participation takes a more expansive view of the methodology available. It uses a justification from first principles or meta-criteria to constitute itself as a knowledge-creator able to assist in the production of regulatory norms for the assessment and management of uncertain risks. The meta-criteria on the menu may be ethico-political (e.g. the common good[16]) – an objective, meta-rationalising principle able to find its way through the competing rationalities of science and social discourse – or objective, deliberative procedures able to deliver coherent regulatory responses by the satisfactory and ongoing resolution of conflicts. The interesting point is that while such theorists tend to take as a given that the rationalities of science, politics, ethics, society, etc. are each distinct from the other, the way in which meta-certainties are used to harmonise the various rationalities is modelled on the rationality of science. This use of meta-certainties enables scientific participation to constitute itself and function. One may say, by way of an imminent critique, that showing that this mode of self-constitution is justified begs the question of how participation itself should be legitimised. If participation is meant to legitimise regulatory responses to risk, what will legitimise acts of participation? A further process of participation to validate the acts of participation in question? Clearly, such a process will just involve an infinite regress and get us nowhere.

The descriptions that scientific participation uses to demarcate its territory are instructive. Whether describing the challenge to be addressed, the discourses capable of dealing with the challenge faced, the participatory model of risk regulation to be used or the required features of the deliberative process itself, the general sense that rationality – and more specifically a scientific rationality – is to be employed, is treated as uncontroversial.

Understanding Uncertain Risks

Andreas Klinke describes the challenge faced in the context of uncertain risks in terms of complexity, uncertainty and ambiguity, all being inextricably intertwined.[17] For him, complexity is about identifying and quantifying causal links. Causal links may be non-linear but are always within the province of science. Where they are highly non-linear or chaotic, complexity turns into indeterminacy, the point at which the limits of analytic modelling are reached (Klinke 2009: 401). Indeterminacy and also ignorance about causal links both create uncertainty which reduces confidence in the estimated cause and effect chain (Klinke 2009: 402). The final component of the challenge faced is ambiguity, denoting the variability of (legitimate) interpretations based on identical observations or data assessments (Klinke 2009: 402).

Klinke uses an empirical, inductive experimental method to establish causal chains that constitute the facts to be regulated. To the extent that we can use a participatory discourse to respond to these facts, it will take the form of a framing of hypotheses which are then contested. These hypotheses are not just those of experts but also include those of the lay public who may introduce more qualitative features like familiarity and notions of the common good or fairness. He assumes that it will always be possible to weigh up these qualitative features and, indeed, to articulate them. In essence, the phenomena that scientific participation has to deal with are chains of cause and effect of varying complexity. What gives a participatory discourse explanatory power and effectiveness is its ability to identify and accommodate these chains. No distinction is made in relation to uncertain systemic risks (and the attendant difficulties of finding an objective stand-point from which to survey them) and uncertain risks within a system where scientific methods may be more capable of delivering – even if only at a future point in time – factual knowledge.

Participation as Method

Ortwin Renn points out that unlike uncertain risks, 'simple' risks require nothing more than the instrumental discourse of the regulator and its enforcement personnel (Renn 2005: 51). Responding to complexity, uncertainty and ambiguity and the possible indeterminacy they carry is more demanding and requires, respectively, epistemological, reflective and participatory discourses. Epistemological discourse should be inspired by different science camps and the participation of experts and knowledge carriers (Renn 2005: 52) across a range of disciplines where necessary. Its aim is to resolve cognitive conflicts in order to produce, as Klinke explains, a homogeneous and consistent definition of and explanation for the phenomenon in question (Klinke 2009: 403). Where the discourse needs the input of both natural and social scientists, additional risk debates and controversies may be anticipated.

Where risks carry a high degree of uncertainty, scientific evaluation is just the starting point for a reflective discourse. The players of this discourse will extend beyond scientists 'to include the main stakeholders in the evaluation process and ask them to find a consensus on the extra margin of safety in which they would be willing to invest in exchange for avoiding potentially catastrophic consequences' (Renn 2005: 52). The aim here is to find the right balance of precaution by assessing the cost of taking the risk against the reward of future benefits. 'Value-tree analysis and other decision-aiding tools [...] can and should be used as structuring instruments within this discourse' (Klinke 2009: 404).

The fourth and final discourse allowing us to respond to the challenge of risks arising from ambiguity, is participatory discourse which is aimed at 'resolving ambiguities and differences about values [by] the subjective weighting of evaluation criteria and an interpretation of the results' (Klinke 2009: 404). The issues covered will include those of fairness, justice, societal change, desired lifestyles and community life.

> The opportunity for resolving [...] conflicting expectations lies in the process of identifying common values, defining options that allow

people to live their own vision of a 'good life' without compromising the vision of others, to find equitable and just distribution rules when it comes to common resources and to activate institutional means for reaching common welfare so all can reap the collective benefits

Renn 2005: 52

The procedures used may include randomly selected juries, voluntary committees and 'consensus conferences'.

The effectiveness of the responsive discourses to the series of causal chains with which they are confronted is measured against norms of homogeneity and consistency. Ultimately, consensus is to be achieved by a risk-reward analysis balancing the discrete elements of science with similarly discrete social values. Some of the super-objective meta-criteria described here are common values, equitable and just distribution rules, and common welfare. The underlying belief is that these criteria will allow us to stitch conflicting demands into a unified whole. Again, we find echoes of a scientific model that aims at achieving an unambiguous delineation of facts which can be coherently analysed and used to produce calculable outcomes within a unified framework.

Scientifying Participation

For Klinke, risk regulation is to be placed within the framework of current developments in governance which occurs within multilevel structures and networks of public and private actors.[18] The top-down approach of traditional risk analysis models is replaced by a horizontally organised structure of functional self-regulation which encompasses both state and non-state actors (Klinke 2009: 405). Within this context the normative elements of the traditional model lose their relevance: new categories for understanding norms need to be established to provide legitimacy and effectiveness. Klinke does not specify what these new categories would look like and where he does refer to legitimacy and effectiveness, they appear to be pretty much the same as you would expect in the traditional model of risk regulation. The rationality and concepts offered by the unitary approach and those on offer in the multilevel discourse of scientific participation overlap to the point where they become virtually identical.

The deliberative approach to be employed in the regulatory model involving participation does not rely on a territorially defined space but on 'functionally differentiated governance' based on fairness and cognitive validity. Its aim is the establishment of the necessary epistemological, reflective and participatory discourses in order to create 'rationally acceptable' risk policy. It attempts to ensure the reconciliation of different demands by the institutionalisation of fair access to risk policy making by dialogue orientated interaction. Institutionalisation raises questions as to how non-state actors are to be selected for participation but its product does not rely on an independent standard;

instead, it embodies the result of cognitive reflection and interpretation within argumentative procedures (Klinke 2009: 408).

Klinke's linking of deliberation to a functionally differentiated governance which creates an embodied product within interpretive and argumentative – and therefore potentially political – discourse, provides a glimpse of how participatory responses may originate outside the framework of scientific rationality. These references are fleeting however and deliberation resumes its scientific guise to become a procedure aimed at the generation of a commonly accepted risk definition, the evaluation of risk consequences on the basis of recognised criteria, and the factual adequacy and common good orientation of the risk policy (Klinke 2009: 408).

Renn reaches a similar place in his discussion of inclusive governance, i.e. governance involving a wide number of participatory bodies. He notes that attention should be paid to the process by which closure of debates (be they final or temporary) is brought forth as well as to the quality of the decision or recommendation that is generated through the closure procedure (Renn 2005: 50). Some of the questions that need to be asked to ensure the quality of the closure process include those checking for the validity of truth claims against commonly agreed standards of validation, and also for a major effort having been made to harmonise different interests and values to find fair and balanced solutions. Again, we are given little direction as to how a commonly agreed standard or a balanced solution might look. We can certainly sense though that we are being encouraged to depend on a common, universal framework which will show itself if only we would deliberate hard enough.

Scientific Participation: Replicating the Risk Analysis Model?

We noted above that a top-down approach and a desire for certainty produced by a scientific reductionism, i.e. a breaking down of complex phenomena into discrete constituents, are key features of the traditional model of risk regulation. Uncertain risks disrupt this methodology and demand a shift from this model. This demand has been answered by a scientific participation which, although it attempts to be multidisciplinary and embed the risk analysis in social processes, may be seen to re-deploy similar strategies.

Top-Down Objectivities

The unitary and linear transmission of power in the traditional model is reconfigured as a search in participatory discourse for a unitary super-objective rationality, a meta-criterion, which will mediate the different, and possibly competing, rationalities of scientific and social discourse. The impulse to seek out this unifying rationality may be seen to arise from the belief that it is the only way in which legitimacy may be preserved and transparency and accountability grounded. The risk that this scientific participation runs is of creating a never-ending circularity which is incapable of determining whether it is participation that legitimises norms or vice versa.

Certainty

Scientific participation works on the basis of the resolution of ambiguities: the creation of an homogeneous and coherent discourse developed from fully articulated and universally agreed definitions. Indeed it sees producing these to be its defining task. Effectiveness is measured by the degree to which the process produces certain outcomes.

Discrete Complexities

Phenomena are modelled on a series of causal links which need a multi-disciplinary investigation in order to be unpicked. While they may be thought of as complex, the underlying presupposition is that the complexity can be managed and uncertainties resolved. Viewing phenomena in this way may better fit the needs of inductive scientific approaches than the socially and physically constituted world in which they are found. There is no real shift in the way in which sovereign, State, regulator, regulatory response and civil society are understood. Instead of joining the dots to connect nodes of power, the aim is to find the interconnections between cause and effect through the efforts of experts from the natural and social sciences.

Like the traditional model of risk analysis, scientific participation depends on a unifying and unified responsiveness in regulatory governance. This tendency to unity is likely to institutionalise its discourse and keep it within the control of experts. Being a participant will be less a right and more of the result of an invitation to join a discourse, the parameters of which are set by experts making objective rationality claims.

A Different Model of Science?

Could scientific participation avoid some of these consequences if a different model of science formed its base? It has been argued that science has tended to adopt a Newtonian approach and search for universal, exception-less laws (Mitchell 2009). Science in this mould will be reductive and seek to base explanations on causes which can be neatly separated from each other. Sandra Mitchell suggests that such a view of science does not take account of complex events which require a revised and expanded epistemology (Mitchell 2009: 3). She offers an outline of such an epistemology which she calls integrative pluralism and which incorporates multiple explanations rather than a single reductive explanation and requires pragmatism instead of absolutism, i.e. 'recognizes there are many ways to accurately, if partially, represent the nature of nature [...]. Which representation 'works' best is dependent [...] on our interests and abilities' (Mitchell 2009: 13). For Mitchell, the policy implications of such a view require us to move from the predict-and-act approach which informs traditional models of risk analysis to a strategy that allows regulatory responses to adapt to evolving changes in knowledge.

A participation that modelled itself on integrative pluralism rather than on a reductive, Newtonian science, would similarly have to be pluralistic and pragmatic. This is not far from the territory which has already been won by the proponents of a participatory model of regulation. Participation provides a forum for a wide variety of views to be expressed and counted (Renn 2005: 50–51). Its justification is precisely a pragmatic one: participation allows regulation to be connected to the community it expects to serve. So the true need of a participatory model of regulation is not to change the scientific model (although that may well be welcome depending on the task at hand), but to accept that the demands for concrete regulatory outcomes from the participation process will potentially lead to a search for super-objective meta-criteria to arbitrate between science and social concerns. Meta-criteria in themselves may appear to be benign but they carry with them the possibility, particularly in cases of uncertain systemic risks, of disabling politics and replacing it with an anodyne participatory process that would have to be guided by pre-determined and super-objective notions of the common good or commonly agreed standards in order to produce results.

Conclusion: Governance and Governmentality

Governance involves the interplay between State and civic society, government and citizens, to produce a combination of hard and soft laws that regulate and allow modern democracies to function. Governance tends to be seen as a good thing: it can, for instance, appear to ensure that laws are relevant or can be seen to limit the arbitrary powers of the State by the input of non-state actors. Participation, with its emphasis on providing a forum for the regulator and the regulated to interact, is central to the risk governance concept and, indeed, effectively amounts to risk governance in practice. Governance and participation both place an emphasis on particular contexts and the need to account for varying needs and multiple interests. It is important then to enquire as to the implications for governance of any implied super-objectivity of the participatory process. A participation so fashioned can make us overstate the availability of meta-criteria – as a means of dealing with competing claims – and hence dissipate the impact of such claims as political initiatives. Often, this manoeuvre is less likely to be a concern where we are dealing with risks within a system given that the system in place will itself establish the common ground between participants and allow a functioning governance to be put in place. Where risks are systemic the search for, or the application of, meta-criteria through a participatory process can be misleading as there will be no available vantage point outside the system to determine their relevance or validity. If this point is ignored, any mode of governance established through participation may be seen to be a displacement of the political process. The very process itself would become a system of regulatory control justified by *faux* meta-criteria rather than, as it is meant to be, a system of regulatory influence.

Once described in these terms, participation and indeed governance can begin to look very much like a form of governmentality (Sørensen and Torfing 2007: 106–8). The habitual practices which create the soft laws of governance are sharpened into hard necessities:

> In governmentality, emphasis would be on how that which appears as necessary is to be understood as assembled together out of available materials, ideas, practices, and so on, in response to a specific understanding of the nature of the problems to be solved
>
> O'Malley 2008: 54

Governmentality was a term introduced by Foucault to denote strategies of control employed by States to empower their political programmes (Foucault 1991). A typical strategy is a process of normalisation whereby techniques are used to establish knowledge and institutional processes which obtain buy-in and enforceability by delivering norms that appear to be allied to individual or collective social ends. Participation may be seen to be one such process which can be massaged by tales of common welfare to deliver the impression of safeguarding a statal–civic alliance.[19] Yet to do so, it must depend on meta-criteria which it will be hard-pressed to justify in terms of its science-based methodology. And without such a justification, participation will always be at risk of a colonisation and enclosure of the common good by an establishment keen to bring it into service for its own ends. The irony is that through participation acting as law-enforcer, these ends will be adopted by the very citizens the establishment needs to control to meet its aims.

Although it may never be quite able to avoid the threat of governmentality, participation need not be modelled in terms of a scientific rationalisation. At the very least, distancing itself from science might allow it to be more politically robust. If the aim of participation is to deal with uncertainties by embedding scientific discourse in a social one, it may be preferable to first examine how social and political bonds operate in order to create socially linked forms of participation. The emphasis on the fact that responsiveness to uncertainty will involve a variety of multi-level social and political choices will allow a model of participation shaped by social bonds criss-crossing, influencing and being influenced by interests within and outside communities. Bonds of community ossify when a scientific participation orchestrates choices by trumpeting its access to super-objectivities and meta-criteria: social bonds fundamentally come about through acts of will, less than through acts of reason. The methods of scientific participation can side-line these bonds by a judicious use of the super-objective.

What participation needs then is a paradigm shift from constitutive rationalities to constitutive communities. This shift will require us to re-conceptualise the notions of trust, accountability, legitimacy and transparency attached to regulation. These notions form the originary cluster of concepts at the core of risk analysis or traditional risk governance and also continue to permeate participatory models of risk governance. A re-examination of these notions will require us to look again at how the spheres of State, non-state and the regulatory response itself, are conceptualised. A finding that these spheres will, at some point or the other, provide intractable resistance to efforts to scientify them, will have a knock-on effect on our conceptualisation of the participatory process and allow us to return it to its political home.

Notes

1 For a useful overview of cases where such an approach has not in fact been adequate, see International Risk Governance Council (2009).
2 Löfstedt and Van Asselt (2008) provides a brief review of work in this area and criticises the IRGC Framework for minimising the importance of social values, highlighted through a process of public participation, in making risk assessments. The Framework however used social values to assess risk management choices.
3 Douglas (2009: 156–74) provides a useful discussion of the importance of moral and cultural values in risk assessment. Andersson, K. (2008: 78–81), cautions against the conversion of public values into quantitative frameworks which could bring the discussion back into the control of scientific experts.
4 It is useful to ask, and a topic for another paper, how the participating individual self is conceptualised. The underlying picture appears to be based on a market model where individuals have choices which they make through a process of self-determination. The fact that something is chosen makes it consumable, legitimises it and disables protest.
5 For an illuminating description of how ideology can masquerade as widely accepted social structures, see Žižek (1989: 33). Žižek writes that 'The fundamental level of ideology, however, is not of an illusion masking the real state of things but that of an (unconscious) fantasy structuring our social reality itself'.
6 See Aven and Renn (2010: 196–97) for examples of such successes.
7 Lee (2010: 818), for instance notes that attempting to neatly separate the technical from the political in risk management can enhance 'the apparent objectivity, inevitability and universality of risk assessment, making it more likely that [.] political decision-makers will seek the safety of expert risk assessment when they explain their decisions'.
8 Dorn (2010: 23–45), suggests that a common response should be avoided as international networking leads to a herding of regulators, thereby increasing systemic risk. Instead, he calls for democratic oversight in order to increase regulatory diversity 'reflecting the twists and turns, indeed the idiosyncrasies of public debate, the unpredictability of populist agenda setting'.
9 A thorough account of the various theories of participatory legislative decision-making, from the neo-liberal to the post-modern, is provided in Renn (2008: 284–306).
10 Discussions focus more on the style of discourse and how to ensure that decision-making is inclusive. See for example Dreyer and Renn (2009).
11 Everson and Vos (2009a). Several essays in this volume rehearse the relevance of participation.
12 Bonzon (2008: 755), notes that commentators have disagreed about the legitimising effect of participation with some arguing that public deliberation only works within an homogeneous polity and others claiming that public deliberation contributes to the emergence of such a polity.
13 See for instance Costas and Žižek (2010).
14 Everson and Vos (2009) provide several instances of this approach.

15 For an example of an explicit link see Klinke (2009). A more implicit link is made by Everson and Vos (2009) in the same volume.
16 For a useful discussion on the role of the idea of a common good in participatory democracy, see Hansen (2007: 249–51).
17 See Klinke (2009) which provides an undiluted example of scientific participation. Some of Klinke's views are heavily influenced by Renn (2005). I have used both studies to outline the key features of scientific participation.
18 Kinke (2009: 405). Renn (2005) too, adopts a similar approach.
19 Flear and Vakulenko (2010) argue that where participation is not oriented toward human rights, it becomes a means of risk managing unruly public perceptions, a legitimating technique rather than a substantive input into governance. See also Power (2007). Also Rothstein, Huber and Gaskell (2006: 91).

References

Agamben, G. *et al.* (2011) *Democracy in what State?* New York/Chichester/West Sussex: Columbia University Press.

Andersson, K. (2008) *Transparency and Accountability in Science and Politics: The Awareness Principle*, Basingstoke: Palgrave Macmillan.

Aven, T. and Renn, O. (2010) *Risk Management and Governance: Concepts, Guidelines and Applications*, Heidelberg/Dordrecht/London/New York: Springer.

Blunkett, D. (2011), 'If I ruled the World', *Prospect*, April issue, p. 4.

Bonzon, Y. (2008) 'Institutionalizing Public Participation in WTO Decision-making: Some Conceptual Hurdles and Avenues', *Journal of International Economic Law*, 11(4): 751–77.

Costas, C. and Žižek, S. (eds) (2010) *The Idea of Communism*, London/New York: Verso.

Defra Consultation (2010) 'Bovine Tuberculosis: the Government's Approach to Tackling the Disease and Consultation on a Badger Control Policy', available online: <http://archive.defra.gov.uk/corporate/consult/tb-control-measures/index.htm> (accessed 3 May 2011).

Dorn, N. (2010) 'The Governance of Securities: Ponzi Finance, Regulatory Convergence, Credit Crunch', *British Journal of Criminology*, 50: 23–45.

Douglas, H.E. (2009) *Science, Policy, and the Value-Free Ideal*, Pittsburg: University of Pittsburgh Press.

Dreyer, M. and Renn, O. (2009) 'A Structured Approach to Participation', in M. Dreyer and O. Renn (eds), *Food Safety Governance*, Berlin/Heidelberg: Springer-Verlag, 111–20.

Everson, M. and Vos, E. (eds) (2009a) *Uncertain Risks Regulated*, Oxford/New York: Routledge-Cavendish.

Everson, M. and Vos, E. (2009b) 'The Scientification of Politics and the Politicisation of Science', in M. Everson and E. Vos (eds), *Uncertain Risks Regulated*, Oxford/New York: Routledge-Cavendish, 1–17.

Flear, M.L. and Vakulenko, A. (2010) 'A Human Rights Perspective on Citizen Participation in the EU's Governance of New Technologies', *Human Rights Review*, 10(4): 661–88.

Foucault, M. (1991) 'Governmentality', in G. Burchell, C. Gordon and P. Miller (eds), *The Foucault Effect: Studies in Governmentality*, Chicago: University of Chicago Press, 87–104.

Hansen, A.D. (2007) 'Governance Networks and Participation', in E. Sørensen and J. Torfing (eds), *Theories of Democratic Network Governance*, Hampshire/New York: Palgrave Macmillan, 249–51.

International Risk Governance Council (2009) *Risk Governance Deficits: An analysis and illustration of the most common deficits in risk governance*, Geneva: IRGC.

Klinke, A. (2009) 'Inclusive Risk Governance through Discourse, Deliberation and Participation', in M. Everson and E. Vos (eds), *Uncertain Risks Regulated*, Oxford/New York: Routledge-Cavendish, 399–413.

Lee, M. (2010) 'Risk and Beyond: EU Regulation of Nanotechnology', *European Law Review*, 35(6): 799–821.

Löfstedt, R. and Van Asselt, M. (2008) 'A Framework for Risk Governance Revisited', in O. Renn and K. Walker (eds), *Global Risk Governance: Concept and Practice Using the IRGC Framework*, The Netherlands, Place: Springer, 77–86.

Mitchell, S.D. (2009) *Unsimple Truths*, Chicago/London: The University of Chicago Press.

O'Malley, P. (2008) 'Governmentality and Risk', in J.O. Zinn (ed.), *Social Theories of Risk and Uncertainty: An Introduction*, Oxford: Blackwell Publishing, 52–75.

Pickard, J. (2011) 'Labour's TB Warning on Badger Cull', *Financial Times*, 20 April.

Power, M. (2007) *Organized Uncertainty: Designing a World of Risk Management*, Oxford: Oxford University Press.

Renn, O. (2005) *Risk Governance. Towards an Integrative Approach. White Paper No. 1 of the International Risk Governance Council*, Geneva: IRGC.

——(2008) *Risk Governance. Coping with Uncertainty in a Complex World*, London/Sterling: Earthscan.

Rothstein, H., Huber, M. and Gaskell, G. (2006) 'A Theory of Risk Colonization: The Spiralling Regulatory Logics of Societal and Institutional Risk', *Economy and Society*, 35: 91–112.

Sørensen, E. and Torfing, J. (2007) 'Theoretical Approaches to Governance Network Failure', in E. Sørensen and J. Torfing (eds), *Theories of Democratic Network Governance*, Hampshire/New York: Palgrave Macmillan, 106–8.

Webler, T. and Renn, O. (1995) 'A Brief Primer on Participation: Philosophy and Practice', in O. Renn, T. Webler and P. Wiedemann (eds) *Fairness and Competence in Citizen Participation: Evaluating Models for Enviromental Discourse*, Dordrecht/Boston/London: Kluwer Academic Publishers, 17–34.

Wilkinson, I. (2010) *Risk, Vulnerability and Everyday Life*, London/New York: Routledge.

Žižek, S. (1989) *The Sublime Object of Ideology*, London/New York: Verso.

Part IV
Synthesis

11 Regulating Innovation, Trade and Uncertain Risks

Marjolein B.A. van Asselt, Tessa Fox, Esther Versluis and Ellen Vos

Setting the Research Agenda

Although it is not often acknowledged in the risk literature, risk regulation is part of the trade sphere. More often than not risk regulation concerns products traded around the globe such as food,[1] food and feed additives,[2] genetically modified organisms (GMOs),[3] pharmaceuticals,[4] children's products[5] and chemicals.[6] Due to globalisation, the international flows of goods have only increased. Do (or might) these traded, often innovative, products pose risks to humans or the environment? And if so, what measures, which may imply trade restrictions, are adequate to deal with these (uncertain) risks? These questions evoke the regulatory activities usually referred to as risk regulation: governmental interference with market and/or other societal processes to control or anticipate potential adverse consequences to safety and health of humans and the environment (compare Kim et al., in this volume, with reference to Hood et al. 2001). Furthermore, many trade conflicts, especially those brought to the World Trade Organization (WTO), are rooted in, or at least associated with, risk concerns. Thus, trade and risk are interrelated in societal and regulatory discourses and decision making.

Nevertheless, as demonstrated in Chapter 1, the topic of balancing between trade and risk receives little scholarly attention in the risk community. This volume aims to highlight and address this deficiency: as the various chapters demonstrate in different ways, the study of the interrelationship between trade and risk is urgently needed. In this volume, we aim to explore what is to be gained, both in scholarly and practical terms, from studying the trade aspects of risk and the risk aspects of trade.

This collection of interdisciplinary papers written by young scholars taken together offers a critical assessment of European risk regulation in a trade context. It is not, however, just about matters in Europe: many of the issues raised are indicative of the challenges involved in balancing trade and risk more generally. Furthermore, in many policy domains, the EU is allegedly setting an international example (see Fox in this volume), which also adds to the broader relevance of the observations in this book. This volume does not provide final answers to critical questions arising when attempting to balance trade and risk, but it does prove that interdisciplinary research in law and social sciences allows for an improved examination of regulatory arrangements. The current volume is not the first in agenda-setting for research collaboration in risk by lawyers and social scientists (e.g. Jasanoff 1990; Fisher, Jones and von Schomberg 2006; Everson and Vos 2009), but it is the first attempt to reveal how young interdisciplinary oriented scholars take up this challenge. It thus offers a unique opportunity to reflect on the added value of integrating legal and social science perspectives.

Integrating Legal and Social Science Perspectives

In this volume, input from legal and social sciences is combined and contrasted in order to provide complementary views on the interplay between trade and risk. The various chapters indicate that, inspired by social science research, legal scholars feel urged to consider the societal dimensions of risk controversies and trade conflicts as well as the socially constructed nature of expertise. Inspired by legal scholarship, social scientists feel invited to investigate 'phenomena usually studied under the auspices of law' (see Motta in this volume, p. 60). Legal scholars remind social scientists to take serious account of legal frameworks and realities. Zurek and Weimer, both trained in law, and Hristova, trained in political sciences, discuss specific EU law and international legal frameworks, including international trade law regimes that shape the legal reality in which risk regulation is embedded. Furthermore, Kim *et al.*, trained in social sciences, systematically describe the legal reality surrounding risk regulation agencies. Although paying attention to legal details, these chapters are easy to read for social scientists and hence provide a useful reference for scholars who want to broach legal frameworks and regimes relevant to balancing trade and risks.

Research Approaches

Most chapters in this volume are informed by empirical study of European risk regulation practices in the international trade context (see Table 11.1). Only the chapter by Rodrigues, on the idea of public participation in risk governance, is a purely theoretically informed chapter. The objects of study and research approaches are quite different over the various chapters. Motta, Ivanova and Van Asselt studied a particular trade conflict. Zurek studied the legal regime in a particular policy domain, while Weimer and Fox analysed the policy dynamics in a particular case and deal with legal frameworks. At the same time, Weimer's assessment of which EU trade restrictions would be accepted at the WTO level is in a way quite comparable to Hristova's endeavour to contrast legal reality with policy practice. Notwithstanding the differences in the object of study, all these authors, implicitly or explicitly, employed a single-case approach.

Table 11.1 Overview of Chapters

Authors	Disciplinary background	Empirical case(s)	Governance level(s)
Part I: The European Union in Context			
Zurek	Law	Food regulation	EU
Weimer	Law	Cloned food	EU
Motta	Sociology	GMOs	WTO, EU
Ivanova & Van Asselt	Political and social sciences / STS	GMOs	WTO, EU
Part II: Risk Regulation in the European Union			
Hristova	Political sciences	GMOs	EU
Navah, Versluis & Van Asselt	Political and social sciences / STS	GMOs	EU
Fox	Political science / STS	Chemicals	EU
Part III: Taking Stock of Policy Fashions			
Kim, Klika & Versluis	Political science	Food, pharmaceuticals and chemicals	EU
Janssen & Van Asselt	Political and social sciences / STS	Food and feed additives and pharmaceuticals	EU
Rodrigues	Political science / philosophy	-	-

Although in different ways, Kim *et al.*, Janssen and Van Asselt, and Navah *et al.* endorse a comparative approach. Kim *et al.* analyse agencies as a specific governance arrangement and systematically compare three different agencies participating in balancing between trade and risk. Janssen and Van Asselt analyse several court cases around various traded products. Navah *et al.* examine political behaviour in 18 voting procedures on GMOs. In doing so, Navah *et al.* point to the need to investigate the politics of risk, a sphere usually ignored by legal scholars and, more surprisingly, also by social scientists interested in risk regulation.

The empirical studies thus vary in whether they investigate a single case in depth or whether they advance a multi-case, comparative approach. Research approaches, furthermore, range from original data collection and interviews to close reading of legal and policy texts and researching other primary and secondary sources relevant to the object of study, and different combinations of these research approaches.

The chapters by Fox (trained in social sciences) and Weimer (trained in law) are interesting to compare as they are very complementary. Both study the policy process embedded in legal regimes by studying a case involving a controversial substance or technology that requires an EU response. Weimer examines the policy process around a category of food products, while Fox discusses the policy process around a particular chemical. Both Fox and Weimer examine how a regulatory response is constructed in situations of scientific uncertainty. Weimer focuses on a broad range of actors in the policy and political realm, with a particular interest in the European Parliament. She also pays systematic attention to the WTO legal framework. In Fox's account the Parliament and the WTO are less prominent but she extends her analysis to the scientific community, the media and the business sector. So taken together, they provide a rich overview of the kind of actors involved in the construction of a regulatory response. At the same time, Weimer and Fox analyse similar dimensions of the challenges in balancing trade and risk. Firstly, uncertainty takes centre stage in both analyses. Both cases are debated by scientists, who test the EU policy makers in order to find a balance between the promise of technological innovation and the accommodation of societal concerns, while dealing with scientific uncertainty. Secondly, experts play a pivotal role in both cases in which scientific inconclusiveness creates demands for different policy answers/frameworks. Weimer primarily analyses and explains the failure to arrive at a legal framework, while Fox aims to anticipate the regulatory response by means of various regulatory scenarios. At the same time, Weimer's assessment of the potential legal consequences of various regulatory choices in the final part of her paper could also be cast in terms of developing regulatory scenarios. The same actually holds for the way Kim *et al.* anticipate the future role of ECHA, the recently established European chemicals agency.

The two chapters illustrate nicely that different methodological approaches can also provide complementary insights. Weimer has, primarily, conducted an in-depth research of legal and policy documents to identify legal and practical constraints in relation to global trade, whereas Fox has conducted interviews and developed scenarios in order to explore available regulatory options. Taken together, these chapters illustrate differences and similarities between a law-based and a social-science-based research approach to risk regulation in a trade context, and in doing so indicate what can be gained from further integration of legal and social sciences.

Consequently, this volume reveals a variety of research approaches which can be used to increase our understanding of balancing between trade and risk. We hope that the portrayal of different research styles provides a basis for the further development of interdisciplinary research methodology that would allow for symmetric analysis of the legal and social science dimensions of risk regulation as part of the trade sphere. The availability of such interdisciplinary research approaches would in turn support systematic comparative research. Although single case studies are definitely important and useful both from a practical and academic perspective, in view of advantages in terms of analytic generalisability,[7] comparative research is vital for empirically informed theory development.[8] In our view, the ultimate ambition should be to prepare for interdisciplinary methical comparison of risk regulation practices in a trade context. Comparison is, for example, possible over different types of risk, over different policy domains and over different legal regimes or trade contexts. In these early days of legal and social science collaboration in risk research, this idea of interdisciplinary comparative research aiming at empirically informed theory development is nothing more than a vision that, hopefully, will materialise in the longer run.

Complementary and Mutually Challenging

This volume illustrates that legally oriented and social-science-oriented analyses on the social reality of trade and risk are both complementary as well as mutually challenging. That is already visible in the chapters of this volume, many of which compare and contrast legal and social sciences. In this synthesis, we explore the ways in which the various chapters are complementary and mutually challenging.

The chapters by Zurek and Weimer can be qualified as law-based reflections on legal regimes, while making use of social sciences literature. Zurek discusses the need for legal frameworks to be embedded in society, while Weimer investigates the impact of the legal order on decision making. In her analysis of EU governance of 'cloned food', Weimer indicates a tension between responding to societal concerns and legal constraints. Hristova, being a political scientist, analyses what legal frameworks are supposed to do and how they play out in actual policy practice. Inspired by social sciences literature, Zurek, Weimer and Hristova try to understand what legal frameworks do in society. Zurek explores what is required to strengthen the societal basis of legal regimes, while Weimer indicates the need to rethink legal constraints to risk regulation. In complementary ways, those authors trained as legal scholars are concerned with the societal robustness of policy making embedded in the current legal regimes: they do not treat the legal frameworks as cast in concrete but as socially constructed sets of agreements that may, or even should, be reformed in view of (changing) societal needs.

The legal scholars' interest in the societal robustness of policy making is similar to the aims of their fellow scholars trained in the social sciences, who in their chapters approach the issue from the other end. In her strong analysis of European GMO regulation, Hristova actually achieves to both explore the legal regimes and the way these are embedded in policy making and reflecting political choices. Motta, Ivanova and Van Asselt, Kim *et al.*, and Janssen and Van Asselt, reflect on legal reality from a social science perspective. Both Motta and Ivanova and Van Asselt use social science concepts and reasoning to examine the famous *EC Biotech* case, an international trade conflict on GMOs brought to the WTO. Kim *et al.* investigate the actual role(s) of risk regulatory agencies, which, as is evidenced by the countless references to agencies in the other chapters, are critical actors both in the legal and policy reality. In the domain of trade and risk, agencies are a critical arrangement around scientific expertise. The focused analysis by Kim *et al.* is complementary as it adds to the analyses in the other chapters. Taken together, these chapters provide a strong, empirically informed basis for reflecting on the expertise arrangements currently in place.

Janssen and Van Asselt analyse several court cases in which the precautionary principle has been invoked. Janssen and Van Asselt identify problematic patterns in how the Court deals with uncertainty and how precautionary measures are reviewed. They argue that a coherent and consistent vision on the relationship between scientific evidence, uncertainty and precaution is urgently needed to resolve the inconsistencies and quandaries persisting. Such a vision is required to safeguard the possibility of legal review of the use of the precautionary principle. Janssen and Van Asselt conclude that case law is most likely not the way towards this much-needed vision. Although not in these words, they argue that problems in the legal realm cannot always be solved there. In some situations, they have to be referred (back) to the policy realm. Also the social science analyses of the *EC Biotech* Case (Motta, Ivanova and Van Asselt) indicate that conundrums encountered in the legal realm require a broader perspective to be understood and addressed. Social sciences are needed to add to the understanding of the difficulties. The authors with a social science background indicate that some seemingly legal issues can or should be (better) managed in the policy realm or in society more generally.

Conversely, Zurek, Weimer and Hristova convincingly demonstrate that the process of risk regulation and the kind of justification employed in that process are decidedly critical in the legal realm. Social scientists with an interest in risk regulation usually focus on the process of risk regulation without paying due attention to the consequences in the legal reality. On the other hand, legal scholars tend to ignore the societal and policy processes that brought about the policy outcomes which are contested in the legal realm. This volume demonstrates that when legal scholars and social scientists join together, the legal impacts of policy processes as well as the policy and societal impacts of legal frameworks become both visible and researchable.

The interdisciplinary contributions in this volume thus demonstrate the importance and the urgency of understanding the interplays between the legal and policy realms when balancing trade and risk. Both the legal and the social science sources, toolkits and bodies of knowledge are needed to examine and explain these interplays in a symmetric and coherent way. Through such genuinely interdisciplinary socio-legal science research, it is possible to identify flaws, or at least problematic imperfections, in the legal regime as well as problematic practices around risk and trade in the policy arena, which tend to be overlooked when the interplays between the legal and policy realms are not considered. The discussion on the regulation of animal cloning is a perfect example (see Weimer).

In our view, in this set of mutually challenging chapters the theoretically informed reflection on the issue of participation in risk regulation by Rodrigues is most radically challenging. He raises a provocative question: 'why do we have this unquestioning consensus around the value of participation as a regulatory ideal in risk governance?' (p. 222). Also in this volume, for example in the chapter of Weimer, the plea for participation resonates. Rodrigues explains the attractiveness of participation as 'a promise of individual engagement accompanied by the warmth of collective solidarity' (p. 224), but '[d]oes participation [.] aid our understanding of and responsiveness to risk?' (p. 225). He warns that participation can obscure the fact that balancing trade and risk requires 'a political response and not a regulatory one' (p. 226). In his view, participation can only contribute to understanding and responding to the political questions pertaining to risk and trade, if it does not, consciously or implicitly, appeal to a super-objective rationality: '[t]he mistake is to believe that participation through, objective deliberative procedures, will lose its clashing political colours and find, instead, a unifying regulatory remedy' (p. 231). Rodrigues argues that accepting responsiveness to uncertainty involves a variety of multi-level social and political choices. It requires us to reconceptualise many critical concepts, such as the notions of trust, accountability, legitimacy and transparency attached to regulation and the spheres of State, non-state and the regulatory response. On such a basis, it might be possible to develop a politically robust model of participation.

Although radically challenging, Rodrigues's plea for returning participation and risk regulation more generally to its political home is at the same time complementary with lines of thoughts in the other chapters. Motta too stresses the political character of risk(y) policies. Weimer observed that the legal realm 'has become the main institutional arena for the political struggles over animal cloning' (p. 40). Navah *et al.* are also deeply concerned that political disagreement around trade and risk is not solved in the political arena, but that it is delegated to, or migrates into, the legal and scientific realms. Like Rodrigues, they plea for re-politicising inherently political decision making and for attending to the political sphere in understanding and improving decision making around trade and risk. Furthermore, all authors in this volume substantiate Rodrigues's claim that uncertainty not only disrupts but fundamentally clashes with the current regulatory model, in which it is attempted to depoliticise and 'scientificise' decisions pertaining to trade and risk.

European Risk Regulation in a Trade Perspective

The EU is entangled in a web of multilateral and bilateral trade agreements and in a complex system of international governance. Risk regulation is part and parcel of this trade setting. The few authors who do address trade in the scholarly risk literature (see Chapter 1 for references) stress the diversity or even inconsistency of regulatory frameworks between trade partners, and in particular the tensions between the EU and the WTO supranational frameworks. Authors in this volume (e.g. Zurek; Weimer; Motta; Ivanova and Van Asselt and Hristova) also discuss the differences between the EU and WTO regulatory frameworks. Not only do the frameworks differ but the underlying rationale of different regulation collide. This volume provides further evidence for the claim that the tensions between regulatory frameworks, including international trade regimes, have serious consequences for risk regulation, with science as a basis for decision making gaining in importance. Informed by the chapters in this book, we think that the importance of researching, understanding and addressing the trade dimensions of European risk regulation cannot be overstated. It is simply impossible to understand European risk regulation without understanding the trade context, including the relevant legal regimes. In the global trade context, the EU has two roles (Zurek): 1) it represents its Member States in negotiating the global regime and it guarantees compliance by the Member States, and; 2) it advances the interests of its Member States before the relevant global institutions. Taking into account the national autonomy of Member States as well as the controversy among Member States, which only increases in a larger and more diversified Europe, this double role of the EU is adding to the regulatory complexity (e.g. Zurek; Ivanova and Van Asselt; Navah *et al.* and Weimer). Furthermore, in practice the EU is often incapable of guaranteeing Member States' compliance (Zurek; see also Versluis 2007; Börzel *et al.* 2010). The transatlantic trade conflict pertaining to GMOs is exemplary for the problematic nature of the dual rule of the EU and the European Commission in particular. So there is not only a tension between global frameworks, such as the WTO, and the EU regulatory framework – which has already been emphasised in earlier papers on risk regulation in trade context – but also tensions due to the double role of the EU, which is rooted in the intergovernmental and multi-level character of this legislature.

In our view, from this volume three critical questions emerge: 1) is risk regulation biased towards spectacular risks? 2) is the societal issue at stake adequately captured by risk? and; 3) may, or should, societal diversity be accommodated and managed in balancing trade, risk and other societal concerns, and how?

A Bias Towards Spectacular Risks

Various authors (for example, Zurek, Weimer, Hristova and Kim *et al.*) argue that risk regulation in the European Union has largely been shaped by post-BSE crisis thinking. This proposition is broadly shared in the literature and it has, for example, been used as a factor in explaining differences between Europe and various trade partners, the United States in particular (see, for example, Jasanoff 2005; Levidow *et al.* 2007; Pollack and Shaffer 2009; Gabbi 2011 and Vogel 2012). Zurek argues that this crisis-thinking has translated into a focus on risks associated with crisis: 'risks connected with bad quality food are not as *spectacular* and explosive as those of disease outbreaks' (emphasis added, p. 18), with the consequence that 'current food-related problems – specifically those connected with nutrition, such as obesity, which in turn contributes to other devastating health consequences [...] and which in practice affect much wider population' (p. 18) are overlooked. Following her choice of words, Zurek warns for a precedence of spectacular risks over mundane and everyday risks in European risk regulation.

This spectacular risk bias is not only visible in the domain of food. Just the name of the EU regime for regulation of risks associated with chemical plants reveals a comparable spectacular risk bias: it is referred to as the Seveso regime, named after the industrial incident in Seveso, Italy, in 1976. Also in this context, proposals for regulatory reform refer to dramatic accidents as their rationale and, like in the domain of food, they are focused on avoiding repetition, i.e. managing yesterday's accidents instead of future risks (Versluis *et al.* 2010). Like Zurek, Versluis *et al.* (2010) argue that other risks are potentially overlooked. Also in other chapters in this volume, it is indicated that spectacular risks might dominate in the risk regulation process. Weimer, for example, indicates that with regard to cloned food the risk assessments were limited to particular species (e.g. cattle and pigs) and, more importantly, do not include food (e.g. milk and meat) obtained from progeny, namely 'offsping (.), where at least one of the parents was a clone' (Weimer, p. 34), while it is to be expected that 'the food products intended for international trade, most likely derive from clone progeny'. Clones are arguably more spectacular than progeny, while progeny are actually more relevant in terms of international trade. Fox in her analysis of the EU regulation of the use of a particular chemical (Bisphenol A, abbreviated to BPA) in children products, observes that the potential health risks of BPA are heavily debated and well assessed, while the potential risks of substitutes are not even discussed and it is even often unknown what is used as substitute with as a consequence that the 'characteristics of the replacement or substituting chemical substances remain in some cases less clear that in the case of BPA since that substance had been subject to several' risk assessments (p. 161). Although not in those terms, Fox's history of BPA suggests that the negative association with DES[9] renders BPA a spectacular risk, which cloaks the unknown and arguably more mundane risks of substitutes.

The question that emerges from this volume is whether or not risk regulation in the European Union is about risks in general or is actually biased towards spectacular risks. Due to this bias, more mundane risks which actually, due to frequency and likelihood, may have greater health, environmental, socio-economic and other impacts, become marginalised in societal and regulatory discourses. The question about such a spectacular risk bias clearly emerges from this volume but it is unclear whether this is unique to Europe and to what extent it is also at least partly shaped by international discourses on risks and trade. Which risks are actually considered in the trade sphere? Are mundane risks, perhaps, not dramatic enough to defend trade restrictions successfully? And to what extent are necessary risk comparisons and risk–risk trade-offs recognised and appraised in balancing trade and risk?[10]

Beyond Risk

Next to the concern that regulation is biased and limited to spectacular risks, various authors in this volume, Zurek and Hristova most explicitly,[11] convincingly argue that risk should be dealt with as just one class of societal concerns. Weimer too emphasises that controversial technology and products not only raise risk concerns, but also ethical and socio-economic questions. Although not literally in those words, their analyses of European food regulation, GMO regulation and the regulation of 'cloned food', remind us of the classical Wildavsky position (1988) that the regulation of innovation is always about balancing potential advantages and disadvantages in the broadest sense of the word. The discourse of risk, which was once liberating as it allowed society to raise concerns pertaining to human health and in a later stage environmental health, can also be narrowing, as it masks other important societal diversity. The question that emerges from the chapters in this volume is whether the controversies in the trade sphere examined can be sufficiently captured in terms of risk, or whether it is actually needed to reframe the issue at stake. Especially Zurek, Hristova and Weimer suggest that 'diversity' would be an interesting candidate for such a reconceptualising attempt. What would be the added value in scholarly terms if risk regulation were to be reconceptualised as a regulatory practice that aims to manage diversity?

The set of considerations around trade and risk is broad. Hristova points to 'divergent levels of risk acceptance about health and the environment, national objectives for the development of certain types of [...] production or for stimulating innovation and research, socio-economic considerations and public attitudes' (p. 112). Navah *et al.* examine the political diversity among (old and new) Member States with regard to GMOs. In the chapters by Fox and Janssen and Van Asselt various examples of different preferences with regard to particular trade products are brought to the fore. Weimer points to divergent ethical and cultural values around scientific innovation and animal welfare, while Zurek touches upon diversity with regard to food, for example in terms of food production methods, consumption traditions and the state of the economy. This diversity has only increased with the 2004 and 2007 enlargements of the European Union. The 12 new members differed significantly from the old Member States and they were also very different from each other in terms of cultural values, consumer preferences, legislative traditions and the stage of economic, social and political development.

This list, compiled from the various chapters, indicates different dimensions of societal diversity: at the level of values (ethics), at the level of (public, social and political) preferences and attitudes, and at the level of socio-economic and historical differences. The notion of diversity is used as a reference to diversity both in the regulatory and the scientific realms, the latter referring to different interpretations of evidence and uncertainty underlying decisions. Furthermore, the various chapters emphasise that diversity refers also to societal diversity *within* the European Union. At the same time they underline that values and preferences for, and other differences between, trade partners are relevant when balancing between trade and risk. So in this context, diversity refers to a range of societal differences on various levels within the European Union and between the European Union and a diversified world. From the perspective of the European Union, diversity has both an internal and external character.

Accommodating Diversity in Balancing Trade and Risk

The third question raised in this volume is how internal and external diversity is or should be accommodated in decision making among trade partners. Various authors in this volume indicate that legal options to accommodate diversity are not often used currently and that where they are used, they are not very helpful in accommodating the heterogeneous character of the EU and its Member States. Zurek, for example, identifies legal barriers that prevent such instruments from being applied, even in cases and contexts that might seem appropriate for acknowledging such diversity. She concludes that these legal instruments are too inflexible to allow for diversity.

Whilst surely some existing regulatory options could be (better-)used to accommodate diversity, most authors in this volume observe that the current ways of balancing uniformity and trade favouring agreements with diversity, and non-economic concerns favouring agreements, are problematic.

Zurek argues that 'a consistent application of unified[12] solutions will inevitably lead to a further disembedding of [...] regulation and [will] distance it from the society it is meant to serve' (p. 29). She argues that satisfying unified regulatory standards may have detrimental consequences for the national economy and society, which is a strong argument in favour of deliberate non-compliance with EU regulation (Krause 2006, cited in Zurek). Hristova asserts that because diversity is not adequately accommodated, individual Member States prefer to deviate from commonly agreed solutions. Both Zurek and Hristova reflect on the proposed amendment of Directive 2001/18 on the deliberate release of GMOs into the environment. This amendment, when accepted, would allow Member States to allow or ban cultivation of GMOs in their country with reference to social, cultural or ethical concerns specific to that Member State. Zurek warns that the EU, 'in order to get away' from international trade–risk conflicts, may actually fuel new internal conflicts and fragmentation of the regulatory framework for its own Internal Market. Hristova concludes that the European Commission in its reform proposals tends to accommodate such 'differentiation', which formerly was referred to as non-compliance. Instead of attempting to accommodate diversity during the decision-making process, so that it is reflected in the policy outcomes, 'the Commission prefers to give Member States the freedom to decide for themselves rather than to try to work out a solution at the EU level'

(Hristova, p. 123). Hristova argues that in this way differentiation is regulated (by accepting what was formerly referred to as non-compliance), but diversity is not accommodated in the decision-making process at the EU level.

While Zurek is concerned about the disembeddedness of the legal regime and non-compliance as a result, Hristova warns that accommodating diversity through differentiation will affect the constitutional characteristics of the EU, as it implies redistributing political authority. In sum, both Zurek and Hristova emphasise the need to accommodate diversity in the decision-making process, in the policy outcomes and in the legal frameworks which shape the policy process. They are both aware that this is far from easy, but '[t]he fact that the issue [accommodating diversity] is problematic […], should not mean that it is better to disregard it' (Zurek, p. 29).

To be sure, in a diversified Europe and in a heterogeneous trade context, it is increasingly difficult to reach negotiated unified solutions anyway. But even in the smaller Europe prior to 2004 it was quite difficult to bring about agreement. Then and now, many of the seemingly commonly agreed policy outcomes are actually technocratic ways of masking the impossibility to arrive at politically negotiated agreements (Hristova; Navah *et al.* and Kim *et al.*). Van Asselt and Vos (2008) qualified this phenomenon as the 'political deficit' in risk regulation. Most authors in this volume, and some explicitly (such as Rodrigues; Navah *et al.* and Motta), are deeply concerned about this political deficit in balancing trade and societal concerns. We may observe that the balance seems to be tilted in favour of trade rather than other societal concerns.

Weimer convincingly demonstrates that it is not only difficult to arrive at politically agreed policy outcomes within an established legal framework, but that it is increasingly difficult to create legal frameworks in the first place. In view of Zurek's and Hristova's pleas for accommodating diversity in legal frameworks, Weimer's scrutiny of the difficulties in developing a legal framework for 'cloned food' is quite telling: she concludes that 'the stronger the pressure is to comply with international trade obligations' (p. 48), the more difficult it is to develop a responsive framework for risk regulation, namely a legal regime that enables accommodating diversity in the decision-making process and outcome.

At this point, it is relevant to turn to Motta's analysis. Motta convincingly argues that economic timing intrudes into political timing: even if it is politically immature or illegitimate to do so, the economic timing obliges political actors to take decisions (see also Navah *et al.*). Motta argues that this implies that the international trade regime has developed from combating protectionism to a neoliberal legal order in which the economy demands that the State regulate, while at the same time setting limits to its interference. In doing so, she points to the need to reconsider the normative basis of international trade law. Synthesising these contributions, we conclude that the way towards responsive risk frameworks enabling accommodation of diversity seems to require a detour through international trade law and normative analysis.

Although the societal and political heterogeneity is obvious, we argue that, so far, scholarly attention to the question how to accommodate and manage diversity in risk regulation in a global trade context has been insufficient. This volume demonstrates that this question is of key importance in view of the ambitions usually associated with risk governance (Van Asselt and Renn 2011). Reframing the issue at stake as accommodating diversity, is in our view a necessary conceptual step ahead, which will hopefully be taken by a critical mass of scholars interested in the legal and social reality surrounding trade and risk.

Rethinking Expertise

Risk regulation concerns governmental interference with the aim of controlling or anticipating potential adverse consequences for health, safety and environment. As these potential effects transcend the capacity of our senses, science is needed to imagine, identify, anticipate and detect risks. This critical dependence on expertise is commonly agreed upon both by risk regulation practitioners and scholars. However, what kind of regulatory arrangements around expertise are adequate is a highly debated issue. What regulatory role should be granted to expertise in balancing trade, risk and other societal concerns? This question, although not necessarily phrased in that way, runs through the most if not all chapters in this volume. Which theoretical and policy-relevant conclusions can be drawn from the observations, conclusions and empirically informed hypotheses compiled in this book?

The first conclusion we draw is that in order to be more specific and precise, the general plea for rethinking expertise[13] should be rephrased into the need to reconsider the regulatory assumptions, principles and arrangements around risk assessment. In the following we therefore synthesise the insights pertaining to the political role and dimensions of risk assessment.

Political Dimensions of Risk Assessment: What, Who and How?

Various authors in this volume point to different political dimensions of risk assessment. What is assessed in the risk assessment? Who is assessing the risks? How is the risk assessment carried out? What counts as risk assessment in the political and legal realms? Although these questions may look technical or institutional, the various authors in this volume convincingly argue they are highly politically significant issues.

What is Assessed?

What is assessed in the risk assessment? The risk assessment approach determines which risk aspects and concerns are assessed. Fox indicates that issues pertaining to the relevance of dosage as a measure of risk and to the interactions of chemicals also raise questions about which risk assessment approach is adequate. A related question is what information to use in the risk assessment. What counts as relevant data? In a multi-actor context, different types of evidence are brought to the fore, and the evaluation of what counts as evidence may differ significantly (Fox). In the first instance, these may all seem to be highly technical, methodological issues. But, taking into account the fact that risk assessment is an institutionalised, obligatory point of passage, those risk concerns that are (not) dealt with in the risk assessment are relevant to the political debate. One example of this is provided by Weimer, who argues that ethical concerns pertaining to progeny (offspring from cloned animals) disappeared on the political stage as a consequence of the risk assessor's decision not to address this risk. With the phrase 'less transparent, but politically significant boundary work', Jasanoff (2005) refers to the role of risk assessment in setting the political agenda. The empirically informed analyses collected in this volume provide some further evidence for the political consequences of seemingly technical choices pertaining to risk assessment approaches. How to accommodate and manage diversity is a key question in balancing trade and risk. Which risk assessment approach is used shapes, at least to a certain extent, whether and which diversity is or may be acknowledged.

Most traded products are regulated on a case-by-case basis, notwithstanding differences between legal regimes and political cultures. Nevertheless, the question 'Risks of what?' is not as obvious as it might seem. Both Fox and Janssen and Van Asselt demonstrate that assessments of a particular chemical are treated in the political and legal realms as indicative of the risks for a particular class of products. An assumed analogy is used to argue in favour of the plausibility of risk: the product assessed is treated as a sample of a broader group of products and its risk assessment is used to support particular, usually restrictive, measures. In this context, it is arguably relevant that the GMO Panel of the European Food Safety Authority (EFSA), responsible for risk assessments of genetically modified food and feed products, always adopted positive opinions on GMOs by unanimity (Hristova; Kim *et al.*). It might be that the GMO Panel experiences difficulty in producing mixed, concerned or even negative opinions on a particular GMO, as they fear that each particular GMO will be treated as a member of a class. As a consequence, consciously or unconsciously, the burden of proof for (potential) adverse effects may be higher than the level of evidence required to issue a positive opinion. Ivanova and van Asselt (p. 99) also refer to this 'class use' of a risk assessment of a particular GMO. They argue that:

> [g]iven the volume of scientific studies examining the effects of GMOs on human health and the environment, it is highly probable that the mere availability of these studies is already interpreted as disqualifying the insufficiency argument, regardless of whether a risk assessment is carried out for the particular GMO.

Janssen and Van Asselt furthermore observed how policy actors attempted to recycle risk assessments, which were once used to satisfy obligations at a particular point in time, in a new risk management round. In one case, this recycling, referred to in the legal realm as 'passive review' was accepted in court while in another case, the court argued that 'any judgements about uncertainty at a present time [...] needs to be informed by new data' (Janssen and Van Asselt, p. 210). This raises the question as to whether the risks are assessed once and for all, or whether uncertainties are interpreted on the basis of the best available knowledge at the time when risk management decisions are taken.

In sum, the authors in this chapter provide further evidence that see-mingly technical and methodological issues are highly important from a political and legal perspective. What is or is not assessed in the risk assess-ment is, at least to a certain extent, setting the agenda and framing the debate. Furthermore, we observe two types of politically relevant forms of ambiguity. Firstly, ambiguity as to the risk assessment used for a single pro-duct or for a class of products, and secondly, ambiguity as to the risk assessment used to consider risks once and for all, or uncertainties in view of the current state of affairs at a particular point in time. The reflections in this book taken together suggest that these ambiguities might invoke politically relevant asymmetries which complicate the balancing between trade and risk.

Who and Whose?

Other politically relevant questions concern who is doing the risk assessment and whose risk assessments count in the regulatory and legal scenes. Kim *et al.* (see also Zurek) argue that the creation of agencies – a development they qualify as an increasing trend – is a response to the idea of risk assess-ment and risk management as being separate activities: risk assessment being undertaken by scientific experts, while actual decisions are taken at the political level. This functional separation is probably the most controversial institutional principle in the risk literature. It has been qualified as impos-sible, counterproductive and undesirable. Also in the chapters of this book, such fundamental criticism resonates (see for example Hristova and Kim *et al.*). Kim *et al.* convincingly demonstrate that in practice the role of agencies is not limited to risk assessment. Agencies play a crucial role in the actual decision making and in risk communication, something that is further identi-fied by other authors in this volume. Although it is assumed or advocated that agencies are the embodiments of the functional separation, in reality risk assessment, risk management and risk communication are interconnected.

What is relevant here is that, notwithstanding the theoretical debate and the empirical evidence, risk assessment is still often presented as an analysis by disinterested experts, who are independent from actors having an interest in either taking, or refraining from, the potential risks. Jasanoff (1990) and others have already convincingly demonstrated that risk assessment is not car-ried out by scientists but by what Jasanoff refers to as 'regulatory science', namely experts who engage in the production of 'serviceable truths'. The analyses in this volume provide further evidence that the politically and legally relevant risk assessments are carried out by such regulatory scientists in agencies or comparable arrangements involving experts on behalf of governments. As Motta (p. 74) eloquently summarises:

> 'knowledge about risks of (.) products is constructed in response to the demands from the political subsystem; it has not originated from the own functioning logic of scientific research'.

Both Fox and Kim *et al.* provide evidence that governments'[14] role as risk assessors is shared with industry. Also Weimer refers to the involvement of industry in the assessment of adverse effects associated with food from cloned animals. Kim *et al.* even argue that industry's input is not only essential but decisive. They observe that 'pharmaceutical companies actually take part in risk assessment' (p. 176). Concerning pharmaceuticals, there is already much exchange between the agency (EMA) and the industry in the pre-submission phase. Also Ivanova and van Asselt refer to the involvement of industry in the risk assessment. They characterise one of the assessments qualified as risk assessment before the WTO as an 'expert opinion [solely] informed by the safety assessment carried out by an interested party' (p. 90). They wonder why the assessments by the Member States, dismissed by the WTO dispute settlement panel, were not used as counterweight to this industry bias. Kim *et al.* contend that in some risk dossiers, the role of the agencies is arguably reduced to reviewing the industry's interested assessments of the risks. Although differences exist between agencies as to how the relationships with industry are arranged, risk assessment is not an activity carried in splendid scientific isolation at a distance from interested parties. We do not have to argue that industry has strong commercial interests in risk regulation. At the same time, the intense involvement of industry in risk assessment is to a certain extent inevitable, if not obligatory or even desirable.

Motta and Janssen and van Asselt refer to the role of the Court in risk assessment. Janssen and van Asselt state that '[t]he Court frames its cases in such a way that it is compelled to enter into a scientific discussion. Although in its wording it prefers to see its role as one of limited review, in practice it is acting as a risk assessor' (p. 208). Motta also observed that the WTO panel engaged in risk assessment. The conclusion which can be drawn from these observations is: 'understanding risk assessment simply as a role of regulatory scientists does not recognize the role played by the industrial applicant and other actors' (Kim *et al.*).

Risk assessment is more adequately portrayed as a platform for interaction between actors in which information concerning potential risks is constructed, shared and pooled. Kim *et al.* suggest that risk assessment is also treated that way by Member States: '[they] seek to reflect their opinions' (p. 181). Kim *et al.* argue that the matter of who is involved in the risk assessment determines whether or not the national opinions are well harmonised in the risk assessment output. If the risk assessment arrangement is such that the national opinions can be harmonised at the level of risk assessment, it is arguably much more difficult to challenge the risk assessment in the decision-making process. Kim *et al.* observe that risk assessment arrangements to hear Member States' scientific concerns and arguments are combined with the procedure to include minority opinions in the final risk assessment. Ivanova and van Asselt convincingly demonstrate how important it is in the legal arena that divergent views be included in the risk assessment to provide legitimate discretion to the executive. Minority opinions are, furthermore, a way to share a broad range of concerns, including differences in scientific interpretations of uncertainty. So Member States' involvement, which sounds horrible to those who prefer to uphold the utopian ideal of risk assessment as an isolated, purely scientific activity, might actually result in opening up the black box of risk assessment, enabling more political weighing of pros and cons in the decision-making phase and allowing for more reflection in the legal realm.

How?

There is no recognised risk assessment approach in the EU legal order (Janssen and Van Asselt, with reference to De Sadeleer 2007). With the authors in this book, we nevertheless would like to argue that legal authorities tend to favour positivistic assessment approaches: the risk is defined as a function of probability and effect, and both dimensions are considered to be calculable on the basis of statistics. As a consequence, risk assessment is past- instead of future-oriented. But the presence of uncertainty requires that the focus should shift 'from the question of probability of the occurrence to a known hazard to the question of defining a potential hazard and its nature in the first place' (Weimer, p. 49; see also Hristova). In such instances, risk assessment should be more future-oriented, which requires an implicit or explicit scenario approach (Kaplan and Garrick 1981; Giddens 1991; Rogers 2001; Van Asselt *et al.* 2010). Whether or not a particular risk dossier is qualified as uncertain should thus have consequences for which risk assessments are accepted in the policy and legal arenas (see also Fox and Weimer). Various authors in this volume indicate that nevertheless in actual regulatory practice, classic past-based assessments of uncertain risks are unquestionably accepted as risk assessment. Fox argues more generally that risk assessments are often 'attempts to suggest scientific conclusiveness' (p. 152), even in situations were uncertainty is obvious, an observation which runs through the whole volume.

How uncertainty is addressed in the risk assessment is critically important in the political and legal arena, as various authors in this volume have indicated (Weimer; Fox; Ivanova and Van Asselt; Janssen and Van Asselt; Rodrigues). As all authors in this volume demonstrate, dealing with uncertainty is a challenge for regulators. But if scientists do not discuss uncertainties, hence mapping the decision situation (Van Dijk *et al.* 2011), the regulators are not adequately informed. It is doubtful that this lack of uncertainty information facilitate them to adequately deal with uncertainty in developing a regulatory, legally viable response.

What Counts?

The risk assessments that count in the regulatory reality are those carried out in governmental expert bodies. Which evaluation of adverse effects is qualified as 'risk assessment' in the legal realm is critical to the ruling in trade conflicts, as the analysis of Ivanova and van Asselt of the *EC Biotech* dispute reveals. The assessments qualified as risk assessment were those conducted by the regulatory science bodies at the national and European level, which served to satisfy the risk assessment obligations. However, the assessments by the Member States to defend their precautionary measures at a later stage were disqualified. Ivanova and Van Asselt point at asymmetries in how risk assessments are treated in the policy and legal domains: some risk assessments are considered much more critically than others. It can be concluded that the formally required risk assessments carried out in the initial policy-making stage are of critical political importance. Janssen and Van Asselt even go one step further and argue that as soon as formal regulatory experts write something down on the risk issue at stake in that policy-making phase, it will be treated and accepted as a risk assessment in the policy and legal realm. 'The fact that experts have been asked to consider the risks seems to be sufficient [to qualify it] as [...] risk assessment' (Janssen and Van Asselt, p. 207). They observe that statements by regulatory experts in which it is explicitly stated that there was too much uncertainty to do a risk assessment, not only qualified as risk assessment in the regulatory process and the legal review, but in one case it even served to argue proof of risk for a different product. Thus, input by regulatory expert bodies charged with doing risk assessment are uncritically accepted as qualified risk assessments, while other attempts to anticipate (uncertain) risks are as easily disqualified. Taking account of the role played by interested parties, this asymmetry is questionable in view of the need to accommodate diversity.

Risk Assessment as a Political Act

The authors in this book, whether trained in law or social sciences, draw attention to the political role of risk assessment. In the current regulatory arrangements, both at the EU and the WTO levels, risk assessment is an obligatory point of passage. It is an institutionalised necessary step before regulators can engage in risk management. Risk assessment is not obligatory for each traded product. This is most explicitly formulated in REACH (the EU framework for the regulations of chemicals) where it is stated that risk assessment is on request (Fox). However, should regulators want to interfere with market and/or societal processes in order to control or anticipate adverse effects, a risk assessment is an obligation: risk management measures need to be preceded by a risk assessment. This procedural requirement also forces legal authorities to determine whether a risk assessment has preceded risk management (Janssen and Van Asselt; Ivanova and Van Asselt; Hristova; Weimer and Motta).

Therefore, Ivanova and van Asselt (p. 97) argue that risk assessment is a political act – as it is an obligatory point of passage, already doing (or refraining from) risk assessment has political consequences:

> Doing a risk assessment is a political act, as it communicates that there is sufficient scientific evidence to do a risk assessment. So as soon as a body [...] considered reputable enough has carried out a risk assessment, it seems no longer possible to satisfy the insufficiency requirement.

This insufficiency requirement is critical in view of risk management possibilities. Ivanova and Van Asselt conclude that as risk assessment is institutionalised as an obligatory point of passage, 'a risk assessment is needed to defend precautionary measures, while at the same time the risk assessment will be used as grounds to argue that the insufficiency condition is not fulfilled' (p. 98). This is most evident at the WTO level, which is also obvious in Weimer's anticipation of the WTO response to precautionary measures pertaining to food from cloned animals. Risk assessment is a political act, as only its existence can already pre-empt risk management.

Although risk assessment is an obligatory point of passage for regulators, the relationship between risk assessment and the actual regulatory response is much debated (see, for example, Ivanova and Van Asselt). In the case of BPA, as examined by Fox, in issuing an emergency ban, the regulators deviated from the risk assessment, which suggested that no change of the existing regulation was needed. In a different way, this also illustrates that risk assessment is a necessary hurdle to take: the regulators issued their emergency ban only after the risk assessment was published. Risk assessment can pre-empt risk management but at the same time it does not necessarily determine regulatory responses either. This is not only visible in the case of BPA. Similar deviations from risk assessments have been observed in GMO regulation (various chapters) and in other cases (Janssen and Van Asselt). It can be argued that such 'deviations' from the risk assessments demonstrate that the actual decisions are taken in the political realm, which would be in accordance with the ideal typical division of responsibilities. However, the inconsistency between risk assessment and risk management is used to contest the regulatory measures, both at the WTO level (Ivanova and Van Asselt; Weimer) and, although to a lesser extent, in European courts (Janssen and Van Asselt). In the legal realm, the discretion of the executive is not necessarily affirmed. In other words, legal authorities do not by definition uphold the regulators' freedom to decide, and hence the freedom to deviate from, the risk assessor's policy advice. This book illustrates in a different way that when institutionalised as obligatory point of passage, any risk assessment is a political act (see also Lee 2012).

These contributions thus make clear that risk assessment is not an apolitical stage. We do not argue that risk assessment is not a valuable activity in balancing between trade and risk, but we have drawn the attention to its political character. That risk assessments amount to political acts is the result of the political nature of risk decisions, the institutional arrangements currently in place and the assessment practices. What is assessed, who is assessing the risk and what counts as risk assessment in the political and legal realities are highly politically relevant questions.

Furthermore, already within the regulatory framework, boundaries have been set on the room for deviation. Kim *et al.* observe that the opinions of the European Medical Authority (EMA) on the risks of pharmaceuticals are not legally binding but that in practice the policy makers do not deviate from them. Should the policy makers want to do, they must provide a detailed explanation. In case of 'scientific and technical doubt', which arguably can be interpreted as any reference to scientific uncertainty, the dossier has to be referred back to the risk assessor. Kim *et al.* conclude that in practice the risk assessment is binding and has a political impact. Hristova and Kim *et al.* argue that this also holds for the risk assessments by the European Food Safety Authority (EFSA) on GMOs. The risk management phase is often reduced to rubber-stamping the risk assessment, even if the political support for the decision is lacking (Navah *et al.*). Or as Hristova phrases it: 'the Commission underutilises its role as risk manager' (p. 123). This all provides further substance for the idea of risk assessment as a political act.

The Politics of Risk Assessment

As the chapters in this volume demonstrate, innovative products in particular involve scientific uncertainty. How can risk assessment be improved to better accommodate diversity in balancing trade and risk? We argue that risk assessment must be recognised as a political act if it is to serve as a valuable stage in accommodating diversity. From the chapters in this volume, we infer two challenges, which could also be read as recommendations for risk regulation practitioners: 1) progressing towards uncertainty tolerance, and; 2) 'relaxing' the risk assessment obligation. Both recommendations are intended to make practicable the idea that 'scientific evaluation is just the starting point for a reflective discourse' (Rodrigues, p. 233), a notion which also runs through all chapters in this volume. We assume that progressing towards uncertainty tolerance as well as 'relaxing' the risk assessment obligation will increase the opportunities in the political and legal realms to be (more) responsive to uncertainty. In our view, responsiveness to uncertainty is necessary to accommodate and manage diversity, but it is not a sufficient condition in itself.

Progressing towards Uncertainty Tolerance

Could it be otherwise? The authors in this volume (e.g. Fox; Weimer and Janssen and Van Asselt) demonstrate that it is possible to provide understandable uncertainty information about technology-related risks in a limited number of words. The notion of 'uncertainty information' is used to indicate that uncertainty is not equal to no knowledge at all. Experts often have knowledge that enables them to indicate which uncertainties are (deemed) important, what the (assumed) sources of uncertainties are, whether or not the uncertainty is in principle reducible, and what interpretations of uncertainty can be considered valid in view of the state of affairs (in other words, whether or nor there are limits to interpretative flexibility). Experts can provide information about uncertainty, which might help non-experts to understand uncertainty and to evaluate the impact of uncertainty on policy decisions as well as the political importance of uncertainty.

Notwithstanding all technical details and the natural scientific knowledge base relevant to the assessment of the risks, in this volume it is clear that legal scholars and social scientists are able to outline the key uncertainties and provide uncertainty information. They succeed in indicating which uncertainties are considered important, by whom, the possibilities to reduce the uncertainties and the limits to interpretative flexibility. Their attempts to disclose uncertainty do not result in seas of uncertainty. They turn out to be able to prioritise uncertainties and put differences, divergences and disputes in the relevant scientific communities into context, for example, by pointing to the history of the academic debate and by revealing the academic credentials of the relevant experts (see, for example, Fox). This enables others to evaluate uncertainty. Fox also describes an industry effort to disclose and share uncertainty information around a contested chemical ingredient.

Providing uncertainty information also, it transpires, is not just a matter of listing the uncertainties, which overwhelms decision makers. A different type of synthesis is required. Van Dijk *et al.* (2011) refer to this new mode of synthesis as 'mapping the decision situation', which they contrast with risk assessment as an attempt to resolve uncertainty among experts through proof of plausibility: a definite answer as to whether the product causes a risk (Van Asselt and Vos 2008). We argue that uncertainty tolerance in risk assessment, and therefore also the systematic provision of uncertainty information, would make it more difficult for policy-makers and authorities to get away with equating uncertainty with risk or questionable constructions of uncertainty in defending precautionary measures (Janssen and Van Asselt).

Consequently, both the young scholars participating in this book endeavour and industry prove that it is possible to provide understandable and policy-relevant uncertainty information. They show how to put the ambition of uncertainty information into practice. Likewise, experts participating in risk assessment should also be able to engage in disclosing uncertainty information. We view that it is a matter of, on the one hand, willingness on behalf of the experts to provide uncertainty information and, on the other hand, willingness on behalf of regulators and other stakeholders to demand and use uncertainty information in the decision-making process.

Informed by their empirical study, various authors in this volume reveal that risk assessment is often approached as an attempt to suggest scientific conclusiveness instead of an endeavour which includes the provision of uncertainty information. Frewer *et al.* (2003) have indicated that experts do not communicate about uncertainty, because they fear politicisation, or the emphasising, amplifying or attenuating of uncertainties as a political strategy. However, this uncertainty intolerant attitude violates the expectations of regulators about risk assessment, which are expressed in internationally agreed guidelines. The Codex Alimentarius Commission (2004), to whom the WTO Panel in the *EC Biotech* case also referred (Ivanova and Van Asselt), explicitly states that 'risk assessment should indicate any [...] uncertainties' (Codex Alimentarius Commission 2004 at para 25, as quoted in Ivanova and Van Asselt), while '[r]isk managers should take into account the uncertainties identified in the risk assessment and implement appropriate measures to manage these uncertainties' (*ibid.* at para 18). It is expected that risk assessors provide uncertainty information, which implies that 'uncertainty intolerant risk assessments are political acts as they do not allow policy makers to decide how to deal with uncertainty' (Ivanova and Van Asselt, p. 88). The Codex Alimentarius Commission is an international, public-standards-setting body that is highly influential because the WTO panels refer to it: if parties comply with the guidelines of the Codex Alimentarius, the WTO uses this, among other things, to assume that the parties involved comply with trade obligations. So in the context of balancing trade and risk, it is highly relevant that the Codex Alimentarius Commission explicitly states that '[t]he responsibility for resolving the impact of uncertainty on the risk management decision lies with the risk manager, not with the risk assessors' (Codex Alimentarius Commission 2004 at para 25, as quoted in Ivanova and Van Asselt). It is highly problematic that experts, in view of uncertainty, treat risk assessment as an attempt to dissolve policy-relevant uncertainties instead of an endeavour that includes the provision of uncertainty information. In this way, such experts are actually deviating from the internationally agreed guidelines outlining their risk assessment role.

The various chapters reveal that uncertainty-tolerant risk assessment is not current practice and risk assessors should be reminded of their responsibility with regard to uncertainty information. There is no mechanism to enforce the Codex Alimentarius standards, so they are not legally binding. In theory, the WTO could argue that uncertainty intolerant risk assessments violate the Codex Alimentarius guidelines, which could have consequences for their ruling in the case. However, as Ivanova and Van Asselt demonstrate, today the presence of uncertainty intolerance in the risk assessment is not considered to be a deviation from the Codex Alimentarius guidelines, while the absence of uncertainty information is understood to mean an absence of uncertainty. It is also quite unlikely that at the WTO level parties would invoke the argument that this violates the Codex Alimentarius: the party defending precautionary measures is suffering from uncertainty intolerant risk assessments produced in their own regulatory system, but they will not themselves argue that their experts deviate from the internationally agreed guidelines pertaining to risk assessment. The plaintiff, the party contesting precautionary measures, actually benefits from uncertainty intolerant risk assessments, as was obvious in the *EC Biotech* case. Thus, it is not given that experts involved in risk assessment will be reminded of their responsibility through WTO case law. They might be prompted by the analyses in this volume but the question remains as to how, within the regulatory systems, experts can be motivated and stimulated to provide uncertainty information, which is after all a non-legally binding but internationally agreed guideline pertaining to risk assessment.

Relaxing Risk Assessment Obligations

In legal frameworks, risk assessment has been approached as the guarantee that measures to protect 'public goods' are not used as disguised discriminatory restrictions to trade. The assumption was that risk assessment could serve as an objective arbiter or as Motta phrased it 'science [is assigned] the role of mediating between [...] economy and politics' (p. 66). Behind this assumption is the reasoning that the logic of science is universal and unifying (e.g. Hristova and Rodrigues). These assumptions should be disqualified as naïve and counterproductive. Risk assessment is a political act which is legally enacted. This should be accepted as a matter of fact. At the same time, we conclude that it would be wise to make risk assessment less important institutionally and legally. Various authors in this volume have highlighted that the obligation to carry out a risk assessment and the ways in which it is currently satisfied and endorsed, make it more difficult to respond to societal concerns and therefore also, to accommodate diversity in decisions pertaining to traded products and technology. The 'risk assessment obligation', which should be read as demanding strong scientific justifications for any restrictive measure, favours the 'scientification' of risk regulation (see Weimer in this volume and Everson and Vos 2009).

In our view, however, risk assessment should not be the sole point of reference in decision-making but should be recognised as one of the elements of regulatory decisions in addition to the 'other legitimate factors' such as social, ethical and political concerns, at the national, European or WTO level. Balancing trade with risks is a political risk in itself, as Motta has nicely pointed out (compare Vogel 2012):

Approving the product opens a path of opportunity but the opportunities may not materialize or they can be accompanied by serious and irreversible damage to health and the environment. Not to approve it impedes the existence of possible – but uncertain – opportunities, for fear of potential damage that, in turn, may not materialize.

Motta, p. 72

She concludes that 'risk management becomes a political risk, because the correctness of the decision can only be evaluated in the future' (p. 73). Actors responsible for the decision making thus have some incentive to inform their decision with risk assessment. Relaxing the obligation would force the actors to be more selective about whether and when additional risk assessment, next to the assessments by industry, is of added value for the decision-making process. We would like to emphasise that our view is not an anti-risk assessment plea. As Jasanoff (1990, p. 1) clearly phrased scientific expertise can 'inject a much needed strain of competence and critical intelligence into a regulatory system that otherwise seems all too vulnerable to the demands of politics'. It also allows a choice to be made between risk assessment prior to the decision and structural monitoring and research when the decision is taken, which would also facilitate learning about uncertain threats to health, safety and the environment over time. Ideally, decision makers would then be better supported to strike a balance between the fear of false positives and false negatives.

Priority is currently granted to concerns pertaining to health, safety and environment. These societal concerns are inscribed in institutional arrangements and legal frameworks. This diminishes the consideration of, and even marginalises, other societal concerns, such as socio-economic and ethical concerns. In other words 'values of high importance are not treated accordingly' (Zurek, p. 23). The risk focus seems to make it more difficult to accommodate and manage diversity, as the broad variety of non-trade concerns has to be disguised as risks. We argue that putting less focus on the scientific justifications of regulatory decisions in terms of risk assessment provides more room in the regulatory and legal realms to treat different concerns on equal footing, hence proving more room for re-embedding decision making in society. In this manner, risk assessment could well develop more easily into uncertainty tolerant assessment of the concerns in the broadest sense. Decision-makers would, hopefully, be better positioned to take inherently political decisions in the political realm 'rather than limiting politics to a scientific anchor, a [way forward] [...] tends to go in the direction of more politics' (Motta, p. 77; see also Weimer for a similar plea and Navah *et al.* for an account of the current political deficit in risk regulation).

It is imaginable that the return of balancing trade and risk and accommodating diversity to its political home would also decrease the resort to the legal arena. Weimer and Motto explicitly observe that legislative procedures have become the main institutional arena for political struggles over contested traded products, which they consider to be problematic. Relaxing risk assessment obligations forces legal authorities to shift their focus from (dis)proving the insufficiency argument to (dis)proving the protectionist argument. Legal authorities have to openly discuss what the arguments in favour of protectionism and what the arguments against protectionism are. This suggestion is consistent with the plea of Winickoff *et al.* (2005) (paraphrased in Motta) 'that anti-discrimination should prevail as a principle of the WTO judicial review and not an interpretation of what constitutes scientific sufficiency' (p. 60). Also, Weimer suggests that focusing on the question of whether risk management decisions lead to a *de facto* discrimination against imports would improve the possibility to be responsive to societal concerns. She argues that legal frameworks already exist which impose the obligation of trade partners not to create unnecessary obstacles to trade. These frameworks are 'recognised for being more generous in recognising the importance of certain values when weighted against the negative effects on trade' (Weimer, p. 47). As in all law courts, in the end the judges must weigh the evidence presented them even when no smoking gun is found. Risk assessment cannot do this job for them.

First Steps

Progressing towards uncertainty tolerance and relaxing risk assessment obligations can be perceived as a first step towards the epistemological change advocated by Rodrigues (p. 236):

> we have to move away from the predict-and-act approach which informs traditional models of risk regulation to a strategy which allows regulatory responses to adapt to evolving changes in knowledge

It can be argued then that we aim 'to gradually open up regulatory decision making to include a wider set of considerations' (Zurek, p. 29). We think that progressing towards uncertainty is supported by relaxing the risk assessment obligations, as that decreases the (perceived or assumed) need to unify scientific plurality. Our emphasis on the limits of risk assessment as the sole source dictating regulatory decisions requires, among other things, the rethinking, rewriting and renegotiating of many legislative provisions from the national up to the WTO level. It also requires institutional reform, both in terms of the mandate of institutions charged with risk assessment and in the culture within and around risk assessment bodies. The fact that the progress towards uncertainty tolerance and relaxing risk assessment obligations are demanding, cannot however mean that it is better not to try. The analyses in this volume unanimously suggest that retaining the status quo or going back to the 'bad old days' (Lee 2010) is not an option. The analyses gathered in this volume indicate that both transformations are necessary steps if we would like to develop towards being able to accommodate diversity in decisions affecting trade and risk.

Conclusions

This volume highlights the need to study trade and risk as being interrelated and it aims to explore what is to be gained from studying the risk aspects of trade and the trade aspects of risk. We do not provide final answers to the critical questions raised in this book but the collection of papers written by young scholars proves that interdisciplinary law and social science research allow for an improved examination of risk regulation in a trade context. Both the legal and social science sources, toolkits and bodies of knowledge are needed to examine and explain the interplays between the legal and the policy realms. The variety and complementarity of the research approaches exposed in this book provide a basis for the further development of interdisciplinary approaches, which in turn would provide a basis for comparative research aiming at empirically informed theory development.

Taken together the collection of chapters offers a critical assessment of European risk regulation in a trade context. But the issues raised are of broader relevance: a bias towards spectacular risks, the need to go beyond risk as it is just a particular class of societal concerns, and the need to reframe the issue at stake in balancing trade and risk as accommodating diversity. The question of the regulatory role of expertise runs through all chapters. We concluded that the regulatory assumptions, principles and arrangements around risk assessment need further reflection. From the empirically informed analyses in the various chapters, we concluded that risk assessment is a political act. Seemingly technical, methodologically issues are politically significant because they shape the extent to which there can be room for diversity in the policy process. Risk assessment should be conceptualised as a platform for interaction between actors in which knowledge (including uncertainty information) concerning potential risks is constructed, shared and pooled. The scientification of risk regulation needs to be offset. In our view, the various chapters in this volume convincingly demonstrate that retaining the status quo is not without significant political and societal risks. The next question is how to improve risk assessment in such a way that it facilitates accommodating diversity while at the same time bringing policy making concerning trade, risk and other societal concerns to its political home. In this synthesis we have identified two challenges that can be read as recommendations for risk regulation in a trade world: 1) progressing towards uncertainty tolerance, and 2) relaxing risk assessment obligations. We are convinced that these steps would gradually help to make innovations in risk regulation in such way that balancing trade and risk may be (re-)embedded in society.

Notes

1 See Zurek (this volume) and Weimer (this volume).
2 Janssen and Van Asselt (this volume) and Kim *et al.* (this volume).

3 Hristova (this volume); Motta (this volume); Ivanova and Van Asselt (this volume); Navah *et al.* (this volume).

4 Kim *et al.* (this volume) and Janssen and Van Asselt (this volume).

5 Fox (this volume).

6 Kim *et al.* (this volume) and Fox (this volume).

7 Analytical generalisability (Yin 1994; Van Asselt 2000) is distinct from statistical generalisability, in which the (statistically representative) number of cases is the basis for conclusions on the level of populations. Analytical generalisability refers to the ambition to defend conclusions on the level of general patterns and phenomena beyond single cases, namely to generalise to theoretical propositions. Although it is not impossible to deduce theoretical propositions from one case, from the perspective of triangulation, comparative case-study research provides a stronger basis for analytical generalisation.

8 With reference to Bovens and 't Hart (1996), we endorse the view that theory and empirics are always intertwined in the scholarly study of social reality. We consider the distinction between 'theoretically informed empirical research' and 'empirically informed theory development' very useful to distinguish different styles of research. The notion empirically informed theory development conveys the ambition that the aim is not to comprehend specific cases, but to develop a more generic understanding of social reality, in this volume the reality of balancing between trade and risk.

9 A chemical used to prevent miscarriage, which actually caused birth defects and other adverse and long-term health effects for so-called DES daughters and their children.

10 It can be argued that the bias towards spectacular risks has similarities with what is in political science literature referred to as 'issue salience'. See, for example, Jones and Baumgartner (2005); Kingdon (1995) and Oppermann and Viehrig (2011).

11 But similar concerns can be traced in other chapters, for example, in Motta; Rodrigues and Navah *et al.* and these concerns are furthermore not at odds with the lines of reasoning in the other chapters, which did not more or less explicitly question the risk frame.

12 Zurek also emphasises that these once 'unified solutions' actually reflect the interests of the old Member States, which may conflict with or even depreciate the interests of new Member States, for which reason they have also been referred to as 'quasi-colonial'. Kim *et al.* provide some evidence that even after the enlargement old Member States are still dominating European decision-making with regard trade and risk. New members are not appointed at powerful positions in the scientific committees of regulatory agencies.

13 Also in previous publications of editors of this book, see for example Van Asselt and Vos (2008).

14 Kim *et al.* detail how within the EU, the EU agencies share the risk assessment role with the so-called National Competent Authorities, although they also observed differences between the agencies with regard to the intensity and the organisation of the interplay between such NCAs and the EU agencies. See also Ivanova and Van Asselt.

References

Börzel, T., Hofmann, T., Panke, D. and Sprungk, C. (2010) 'Obstinate and Inefficient: Why Member States do not Comply with European Law', *Comparative Political Studies*, 43(11): 1363–90.

Bovens, M. and 't Hart. P. (1996) *Understanding Policy Fiascoes*, New Brunswick, USA: Transaction Publishers.

De Sadeleer, N. (2007) 'The Precautionary Principle in European Community Health and Environment Law: Sword or Shield for the Nordic Countries?', in N. De Sadeleer (ed.), *Implementing the Precautionary Principle. Approaches from the Nordic Countries, EU and USA*, Earthscan: London.

Everson, M. and Vos, E. (eds) (2009a) *Uncertain Risks Regulated in National, European and International Contexts*, Abingdon UK and New York USA: Routledge-Cavendish.

Everson, M. and Vos, E.I.L. (2009b) 'The Scientification of Politics and the Politicisation of Science', in M. Everson and E. Vos (eds), *Uncertain Risks Regulated*, London: Routledge/Cavendish Publishing, 1–17.

Fisher, E., Jones, J. and von Schomberg, R. (eds) (2006) *Implementing The Precautionary Principle Perspectives and Prospects*, Cheltenham: Edward Elgar.

Frewer, L.J., Hunt, S., Brennan, M., Kuznesof, S., Ness, M. and Ritson, C. (2003) 'The Views of Scientific Experts on how the Public Conceptualize Uncertainty', *Journal of Risk Research*, 2(6): 75–85.

Gabbi, S. (2011) 'Independent Scientific Advice: Comparing Policies on Conflicts of Interest in the EU and the US', *European Journal of Risk Regulation*, 2: 213–26.

Giddens, A. (1991) *Modernity and Identity: Self and Society in the Late Modern Age*, Stanford, CA: Stanford University Press.

Hood, C., Rothstein, H. and Baldwin, R. (2001) *The Government of Risk: Understanding Risk Regulation Regimes*, Oxford: Oxford University Press.

Jasanoff, S (1990) *The Fifth Branch: Science Advisers as Policymakers*, Cambridge, MA/London: Harvard University Press.

Jasanoff, S. (2005) *Designs on Nature: Science and Democracy in Europe and the United States*, Princeton: Princeton University Press.

Jones, B. and Baumgartner, F. (2005) *The Politics of Attention. How Government Prioritizes Problems*, Chicago: The University of Chicago Press.

Kaplan, S and Garrick, B.J. (1981) 'On the Quantitative Definition of Risk', *Risk Analysis*, 1(1): 11–27.

Kingdon, J.W. (1995) *Agendas, Alternatives, and Public Policies*, New York: Harper Collins.

Krause, K. (2006) 'European Union Directives and Poland: A Case Study', *University of Pennsylvania Journal of International Economic Law*, 27: 155–204.

Lee, M. (2012) 'Beyond Safety? The Broadening Scope of Risk Regulation', in *Current Legal Problems 2009*, Oxford: Oxford University Press, 242–85.

Levidow, L., Murphy, J. and Carr, S. (2007) 'Recasting Substantial Equivalence: Transatlantic Governance of GM Food', *Science, Technology and Human Values*, 32(1): 26–64.

Oppermann, K. and Viehrig, H. (eds) (2011) *Issue Salience in International Politics*, London: Routledge.

Pollack, M.A. and Shaffer, G.C. (2009) *When Cooperation Fails: The International Law and Politics of Genetically Modified Foods*, New York: Oxford University Press.

Rogers, M.D. (2001) 'Scientific and Technological Uncertainty, the Precautionary Principle, Scenarios and Risk Management', *Journal of Risk Research*, 4(1): 1–15.

Van Asselt, M.B.A. (2000) *Perspectives on Uncertainty and Risk: The PRIMA Approach to Decision Support*, Dordrecht: Kluwer.

Van Asselt, M.B.A. and Renn, O. (2011) 'Risk Governance', *Journal of Risk Research*, Special Issue 'Uncertainty, Precaution and Risk Governance', 14(4): 431–49.

Van Asselt, M.B.A. and Vos, E. (2008) 'Wrestling with Uncertain Risks: EU Regulation of GMOs and the Uncertainty Paradox', *Journal of Risk Research*, 11: 281–300.

Van Asselt, M.B.A., Van 't Klooster, S.A., Van Notten, P.W.F. and Smits, L.A. (2010) *Foresight in Action: Developing Policy-oriented Scenarios*, London/Washington DC: Earthscan.

Van Dijk, H., Van Rongen, E., Eggermont, G., Lebret, E., Bijker, W. and Timmermans, D. (2011) 'The Role of Scientific Advisory Bodies in Precaution-based Risk Governance Illustrated with the Issue of Uncertain Health Effects of Electromagnetic Fields', *Journal of Risk Research*, Special Issue 'Uncertainty, Precaution and Risk Management', 14(4): 451–66.

Versluis, E., Van Asselt, M.B.A., Fox, T. and Hommels, A. (2010) 'The EU Seveso Regime in Practice: From Uncertainty Blindness to Uncertainty Tolerance', *Journal of Hazardous Materials*, 184: 627–31.

Versluis, E. (2007) 'Even Rules, Uneven Practices: Opening the 'Black Box' of EU Law in Action', *West European Politics*, 30(1): 50–67.

Vogel, D. (2012) *The Politics of Precaution: Regulating Health, Safety, and Environmental Risks in Europe and the United States*, Princeton: Princeton University Press.

Wildavsky, A. (1988) *Searching for Safety*, New Brunswick: Transaction.

Winickoff, D., Jasanoff, S., Busch, L., Grove-White, R. and Wynne, B. (2005) 'Adjudicating the GM Food Wars: Science, Risk and Democracy in World Trade Law', *The Yale Journal of International Law*, 30: 81–123.

Yin, R.K. (1994) *Case Study Research: Design and Methods*, Thousand Oaks, CA: Sage Publications.

Wynne, B. (2001) 'Creating Public Alienation: Expert Cultures of Risk and Ethics', *Science as Culture*, 10(4): 445–81.

Index